CO-CREATE A MEANINGFUL LIFE
AND DEEP CONNECTION
WITH ANIMALS

Intuitive Animal Communication

MICHAEL R. BURKE

HAY HOUSE
Carlsbad, California • New York City
London • Sydney • New Delhi

Published in the United Kingdom by:
Hay House UK Ltd, The Sixth Floor, Watson House,
54 Baker Street, London W1U 7BU
Tel: +44 (0)20 3927 7290; www.hayhouse.co.uk

Published in the United States of America by:
Hay House LLC, PO Box 5100, Carlsbad, CA 92018-5100
Tel: (1) 760 431 7695 or (800) 654 5126; www.hayhouse.com

Published in Australia by:
Hay House Australia Publishing Pty Ltd,
18/36 Ralph St, Alexandria NSW 2015
Tel: (61) 2 9669 4299; www.hayhouse.com.au

Published in India by:
Hay House Publishers India, Muskaan Complex,
Plot No.3, B-2, Vasant Kunj, New Delhi 110 070
Tel: (91) 11 4176 1620; www.hayhouse.co.in

A catalogue record for this book is available from the British Library.

Tradepaper ISBN: 978-1-83782-083-2
E-book ISBN: 978-1-4019-7530-2
Audiobook ISBN: 978-1-4019-7531-9

Cover design: The Book Designers
Interior design: Nick C. Welch

This product uses responsibly sourced papers and/or recycled materials.
For more information, see www.hayhouse.co.uk.

Printed and bound by CPI Group (UK) Ltd, Croydon, CR0 4YY

CONTENTS

Introduction vii

Part I: Creating a Foundation for Your Own Energy and Intuition

Chapter 1: Power in the Present Moment 3

Chapter 2: Setting Intentions and Cultivating Awareness 21

Chapter 3: The Mind, Heart, Spirit, Intuition Connection 35

Chapter 4: Becoming a Stronger Receiver 63

Part II: Processes, Tools, and Techniques

Chapter 5: Understanding Animal Communication 81

Chapter 6: Establishing Your Internal Database and Ritual 95

Chapter 7: Techniques for Communication 111

Chapter 8: Animals in the Afterlife, Spirit Animals, and Power Animals 129

Chapter 9: Beyond the Basics 143

Part III: Putting It All into Practice

Chapter 10: Building Upon Your Foundation 163

Chapter 11: Strengthening and Elevating Your Animal Relationships 179

Chapter 12: Co-creating with Animals 195

Afterword 217

References 219

Acknowledgments 221

About the Author 223

INTRODUCTION

While everyone has a different story when it comes to pets and other animals, I feel safe in saying there are few people whose lives haven't been touched by a creature in some way. Though as humans, we have done a lot to separate our lives from the lives of animals, we still share our world with them and, on some level, will always be connected to them. But that connection can go much deeper—it can become communication. If you've ever connected with a dog, caught the gaze of a wild deer, felt completely understood by a dolphin, or shared a quiet moment with a squirrel while fervently wishing you could understand what they were thinking, this book is for you.

Simply put, I define *intuitive animal communication* as exchanging information with animals on an energetic and telepathic level. This might come about as visual images in your mind's eye, hearing a voice or sounds, emotional feelings or other bodily sensations, just deeply and instantaneously knowing a truth, or another way.

You don't need any background in the exercises I describe or any proven psychic abilities to begin developing your practice. Intuitive communication is just that—intuitive, instinctive, innate—and everyone has an aptitude for this skill. Where most people get stuck is in developing the skill; just like learning to roller-skate or play the trumpet, even if you have talent, you still have to put in the work. That's where this book comes in. My process begins with understanding and strengthening your unique intuitive skills and then combines that practice with techniques for animal communication.

So many clients come to me wanting to be able to understand their pets better and communicate with them on a deeper level, whether to help with behavioral issues or just forge stronger connections. This desire to connect might also extend beyond our pets to the wild animals with whom we share our space. This is all possible, and it can open doors to a new and more integrated way of experiencing life! It sure has for me.

My Story

As far back as I can remember, I have felt a profound connection with animals, and it all began with Cory the collie. Cory lived with the Weisses, just behind my family's home in the suburbs of Chicago where I grew up. Mr. Weiss was a vibrant older man in his seventies with a short and stocky build, rosy cheeks, squinting eyes, and a beaming, joyful smile. He wore a newsboy cap and Chicago Bears jacket with a cigar perpetually in his mouth, and Cory was always by his side. Even before I could walk, I was drawn to them. My dad used to tell me how I would see Cory across the yard and reach out toward him with my chubby baby arms, wanting to make contact. He became friendly with Mr. Weiss and would often take me over to their house to visit. I still hold on to vivid memories and feelings of Cory's presence—strong, wise, patient. I felt safe with Cory.

Cory was even the one who helped me get up on my feet and walk. I would hold on to him as I stood up, and for those first steps, he was there. I could not talk yet, but we did not need to speak. We communicated through energy, and I felt him offering me a space of safety and guidance. I was fascinated by him, practically worshipping him, and I felt an inner knowing that he was supporting me with calm and balanced energy as I engaged with the world.

I am the youngest (by eight years) of six kids born into an Irish Catholic family—three boys and three girls. By the time I came into the family, my siblings were older, as were my parents, and there was a strong dynamic already in place for me to learn from and observe, with each family member serving as a different kind of teacher.

Because of our Catholic upbringing, we were taught to be very aware of our connection to a higher power and things beyond our usual perception. We were taught to pray, communicate with our angels and saints (I was named after Saint Michael), and be mindful of our thoughts and words. My family would gather around the dinner table and discuss our mystical and paranormal experiences—sensing energies, getting intuitive feelings, dreams, experiences with deceased relatives, the sounds of ghosts in our attic, and the "vibes" we would get in a variety of forms. It was always in good fun, while at the same time very real and serious. Everyone loved to explore the mystical and unknown, and we all could not wait to share our experiences because we knew the whole family appreciated them and could relate. This openness to all things related to energy, intuition, the afterlife, and spirituality created an environment where I could explore my intuition without limits.

I was always a sensitive and empathic kid. My mom later told me that it was as if I were a sponge that picked up on everything in my environment and then reacted through my emotions. She said that I was constantly and quietly taking everything in, not just watching but absorbing my surroundings.

Due to my bond with Cory the collie, I was always begging my dad for my own dog. He finally caved in when I was nine years old, and we ended up with an Irish terrier puppy because, as my dad said, we were "a hundred percent Irishmen!" Corky was very stereotypically Irish: he was strong-willed, hotheaded, stubborn, and smart enough to know that he did not need to listen. During obedience classes, Corky would not follow the cues. My dad was adamant about being the one to train Corky, but he was also a very busy attorney who worked long hours. I was at an age when I was busy with school, friends, sports, and other activities. As such, Corky was usually confined to his outdoor dog run or limited space inside the house (he could not be trusted alone). He was not fixed, and he was never house-trained. I walked Corky a few times each day, but I recognized it was not enough. I understood Corky's willfulness and felt from him that he needed more fulfillment, yet I did not know how to offer that. I told my parents that Corky needed more guidance, that we needed more guidance, and that he was just misunderstood. But my parents had many other distractions and did not prioritize Corky's emotional well-being.

Corky was always well taken care of, but throughout his fifteen years, I always felt he deserved more than we offered him.

My family's experience with Corky—the challenge of living with a difficult dog, and of everyone in the household not being on the same page—sparked something within me. It ignited a desire to learn how I could help people like my parents see life from the dog's perspective and help dogs like Corky live a more fulfilled and well-rounded life. I was also beginning to recognize how often I was intuitively tuning in to animals and humans. In doing so, I was feeling the communication gaps that needed to be filled so animals and humans could understand each other better.

During my childhood, I also learned that I could communicate with other animals besides dogs, and they didn't even have to be nearby. When I was 10 or 11, I had several nights of recurring dreams of a cat behind a window, calling out that he was trapped and needed help. I shared my dreams with my dad and told him I felt there may be a cat trapped somewhere nearby. He listened but felt there was not much we could do aside from keeping our eyes and ears peeled.

A few days later, we started to see flyers posted around the neighborhood about a lost cat, and my dad and I were immediately on the case. This was the cat from my dreams! I just knew it. On our walks with Corky, we began to investigate more closely in people's yards. It was not long before we heard a meowing very loud from a neighbor's garage, a neighbor that we knew did not have a cat. We called the phone number on the lost cat flyers and told them we heard unusual cat sounds at the house, providing the address. The guardian of the lost cat asked the homeowner to check their garage, and there he was. It turned out that the housekeeper was not happy that the cat was using their flower garden as a litter box, and she locked him in the garage. The homeowner had no idea. He was very old and could barely walk or hear, and he was completely shocked to discover the cat in his garage. The cat was okay and safely returned to his family. This experience was a huge breakthrough for me in recognizing my intuitive capabilities.

In my high school years, I began strengthening my intuition and learning more about personal development, metaphysics, and all things related to spirituality. My sisters and brothers were all also interested in these topics, and we had bookshelves full of books on astrology, tarot, meditation, extra sensory perception (ESP), past lives, out-of-body experiences, Reiki and energy healing, saints, angels, spirit guides, the law of attraction, manifestation, you name it. My sisters would go see intuitives and psychics, and then play me the cassette tape recordings of their readings. I loved soaking it all in because I then felt validated in the experiences I was having.

Though I had dabbled in all kinds of esoteric knowledge, I began developing a serious spiritual practice at 17, when I was stressed out and preparing for college entrance exams. In my sister's room, I found an audio cassette tape of guided meditations by Shakti Gawain called *Creative Visualization* and gave it a try. I was hooked after one listen. It was like a miracle cure for releasing stress and tension, and the experience was like rocket fuel for my lifelong study of meditation. I also began to realize that my sensitivities were a strength, and I could cultivate and utilize them to help others.

I went off to college in Boulder, Colorado, to study environmental studies and conservation. Living in beautiful Colorado opened me up even more to my intuitive abilities. I felt that the sun, mountains, snow, and crisp air amplified my intuitive senses. During this time, I also recognized the importance of my health and maintaining balance so I could be a stronger receiver, communicator, and support for myself and others. I studied nutrition and became more mindful of the energetics of food, how food affected me, how some foods muted my intuition and energy while others amplified it. I also began to exercise regularly because when I was not as active, I felt very blocked and energetically stifled.

I began to do tarot card readings for people for fun. I seemed to have a knack for it and encountered more and more people who wanted intuitive guidance. One of my friends once said to me, "What is it with you? Everywhere we go, you are like a magnet for people who tell you their life story five minutes after meeting you! Next thing you know, they are going to give you their social security number!" I, too, was aware of how people were drawn to me, and after some time realized that this was my gift, that I had an opportunity to help people by listening and holding space for them. When I surrendered to this perspective, things started to open up even more for me.

After graduation, I could not find any job opportunities related to environmental work that would provide a solid living, so I began a career in advertising sales, something that two of my sisters had found success in. My career began to take off, but deep inside I felt that something was not right. I did not feel aligned with my true path, and ad sales was such an intense, cutthroat world—it was draining my soul. On the side, I continued to learn more about all things related to intuition and energy. I took countless workshops and certifications. I was growing and sifting through what resonated and did not resonate with me. It was also during this time that my dog Cooper came into my life.

On a beautiful spring Sunday, I was taking a walk after brunch when there she was, in the window of a storefront in Chicago where an adoption fair was taking place. Cooper was a Collie–German Shepherd mix, just seven and a half weeks old. Although I had no plans to adopt a dog that day, she was the one; I instantly knew the Universe had brought us together. Cooper was my soulmate. She changed my life forever.

Cooper was a very sensitive dog. She did not like strangers, was very reactive, and had a strong herding instinct—even on walks if the street was crowded. I often said Cooper was like a cat: everything was on her terms. She was only affectionate when she wanted to be, and when she wanted her way, she was very clear about it. She was very challenging but also so loving. Our bond went beyond love, beyond me trying to analyze her behavior. I felt her at an instinctual level, and I felt her feeling me in that way too. We could look at each other and communicate effectively with no words. People were always mesmerized by our bond. Cooper helped me build my confidence in understanding what animals are thinking and feeling.

I learned from Cooper how important it is to be aware of my energy. She was extremely sensitive, so when I was tense, she would be tense. If I was emotional, she would feel unsettled. If I was distracted, she was insecure, mirroring back to me my imbalanced energy and making it clear when I was out of alignment. This was such a gift because she challenged me to be calmer and more aware of what I was projecting, practice what I preached, and be more present in the moment. I began to use the meditation practices I had learned with her, developing my own versions that supported and uplifted our communication. She showed me how powerful meditation was, as the more (and more consistently) I practiced, the more our connection and relationship improved. She gave me opportunities every day to see the results of meditation in improving both my energy and, subsequently, her behavior.

Through word of mouth, people who needed help and were interested in alternative spirituality began to find me. They wanted support, someone to listen to them, and intuitive guidance for their personal journeys. I created a side business offering readings and sessions, and it took off. I was very much in the closet about it, though, because I did not want anything jeopardizing my career.

Knowing my love, connection, and experience with animals, repeat clients began asking me for help with their pets. I began to offer intuitive reading of pets, helping people understand what their pets were thinking and feeling. It often felt like I was a therapist helping to clear the communication gaps between people and their pets. Often, people would come to me with behavior challenges, and I would tune in to the animal and say to the person, "This is what your animal is saying and feeling, but you really just need a dog trainer." It dawned on me that, instead of referring them out to a trainer, I could add training and behavior consulting to my tool kit.

I began studying all aspects of animal behavior. I attended the Animal Behavior College and became certified as a dog trainer and behaviorist. I apprenticed with other behaviorists and trainers to strengthen my handling experience and observe different approaches to challenges people faced with their pets. I could feel the foundation of my work and offerings becoming stronger, and a momentum building in the variety of ways I could help people and their pets. When I trusted my inner guidance and paid attention to the signs all around me, I began recognizing that great things came together and moved me forward on my path.

My side business in animal communication continued to expand and my day job in advertising sales was fading, but I wasn't sure if I was ready to take that leap of faith and quit. But then the company I worked for was acquired by a larger organization, and I was laid off—how's that for a sign from the Universe? I felt empowered to seize the moment and do what I wanted to do.

It was then that my inner guidance told me to look up Cesar Millan, known as the Dog Whisperer. I learned that he created a ranch called the Dog Psychology Center not far from where I lived in Southern California. Without giving it a second thought, I signed up for a workshop, which turned out to be another turning point. I could relate to Cesar's energetic connection with animals and his ability to help dogs and their families overcome behavior challenges. After completing the workshop, I began to volunteer at the Dog Psychology Center on weekends, assisting in training classes and longer workshops, and eventually becoming a trainer. My role on Cesar's team grew into the meditation director for his Training Cesar's Way program, where people travel from all over the world to attend five-day workshops focused on dog behavior and training. I created a curriculum on mindfulness, meditation, energy, manifesting, and personal development to help participants create calm and balance in their lives to improve their connection with animals.

In my work as an animal communicator, behavior consultant, and intuitive coach who connects with thousands of people and animals all over the world, I see that people want and need to live a more mindful and miraculous life with animals. People I speak with are hoping to get inside their pet's mind to understand what they're thinking and feeling, learn how to best support them, and improve their relationship and their lives together. They need help balancing and calming their thoughts and emotions that they know are affecting their pets. They also want to connect with beloved pets in the afterlife or feel more united with nature and their wild animal neighbors. The basis for this kind of life is proactive and consistent animal communication—and it will open your mind to joy you've never dreamed of.

There are a lot of factors that have led us to be at a set point of imbalance, both personally and globally. During the COVID-19 pandemic, more people than ever adopted pets for companionship only to quickly realize they did not know how to properly take care of them. With emotions running high and all of the pandemic-related imbalances due to trauma, stress, fear, and sadness, connection with and care for pets were further compromised. This has needlessly led to an influx of aggravating pet behavioral issues, people who are stuck at home with a pet they can't leave alone, or sadly, pets that end up in shelters.

Our planet is also in the throes of ecological crisis, and our battle against climate change and deforestation that destroys crucial habitats is a battle that includes wild animals. There has also been a huge upswing of interest in spirituality not connected to organized religion. People are interested in the New Thought movement, intuition, energy, and ideas and concepts beyond traditional thinking and perception. This openness has led people to consider that there is more out there than just what we can see, and they desperately want to connect with it.

Various wonderful books and programs teach animal communication. Others teach mindfulness and meditation. There are also books on animal training and behavior, manifestation, energy, and personal development. But I couldn't find anything that combined all of these topics into a cohesive, holistic, and well-rounded system that truly supports and enhances our understanding, communication, and connection with animals. Recognizing a need, I integrated my expertise on all these topics into a video program in 2021 called Intuitive Animal Communication Learning and Development. That course was the inspiration for this book.

How to Use This Book

I have brought together all of my signature processes, tools, and techniques in this book to help you strengthen your abilities to understand what your pets are thinking and feeling so you can better support them, and help you and your animals feel calmer, more balanced, and fulfilled as you co-create a joyful life together.

The book is divided into three parts. Part I is where we put on our own oxygen masks first, exploring why it's important to cultivate calm, confident, and clear energy before interacting with animals. You will learn about journaling and mindfulness, try out some meditative techniques to quiet the mind and tune in to your inner guidance, and practice recognizing and interpreting energy in all forms.

Part II is all about animal communication. While I know this is what you're here for, please don't just skip right to this part! The information and practices in Part I are essential tools that will set you up for success. Here, we'll dive into what intuitive animal communication is, create an internal database of signs and symbols, and establish a process that is right for you. Then we will get into the specific techniques, including how to communicate remotely, call upon animals in the afterlife, work with wild and domesticated animals, contact lost animals, and much more.

Part III focuses on putting it all into practice and stepping out into the world. At this point, you will have an established practice for yourself and can begin making animal communication a lifestyle—more second nature and instinctual, just like communicating with other humans (well, most of the time!). Here, we'll discuss commitment, maintaining energetic boundaries, managing expectations, strengthening your animal relationships, and co-creation.

Throughout the chapters are many exercises that, though they are designed to strengthen your attunement for animal communication, will also benefit your personal growth and human relationships. Everyone is different and not every technique resonates with each person. You may feel that some practices are beneficial for a while and then want to switch it up and try something new. Whatever practices you choose, I encourage you to give yourself some time to adjust to them, as the benefits develop with consistency. Also included in each chapter are journal reflections to offer support in further absorbing the information. This path is a commitment, but it's a joyful one.

With my guidance, you'll gain tools to communicate with animals, but you'll also learn to live intentionally in the present moment, welcoming in more love, joy, compassion, gratitude, connection, and support for yourself, other humans, animals, and the planet we all call home.

Let's begin our journey together!

Part I

CREATING A FOUNDATION FOR YOUR OWN ENERGY AND INTUITION

POWER IN THE PRESENT MOMENT

At the beginning of our journey to enhance our communication, connection, and relationship with animals, it's important to first cultivate a solid foundation in understanding our own energy and intuition and how we engage with the world.

As humans, we spend a lot of time in our heads, focused on the past or the future. This is a defense mechanism—one that was once crucial for our survival as it kept us cautious. But we can easily become stuck in these modes of thinking, making fear-based choices due to past experiences or being so focused on achieving our goals, and what might or might not happen, that we have trouble taking action. Because our attention and awareness are often behind or ahead of us, we also miss out on what is happening in the present and all of the gifts, options, and opportunities available to us right now.

When we are present, we can notice and appreciate the energetic subtleties and beauty each moment has to offer. For example, you may not notice the magic of a pink rose blooming despite it being February if you are looking at your phone while walking your dog. By simply opening up all your senses during any activity, you make yourself available to the fulfilling experience of appreciating whatever beauty you might encounter. In turn, this may inspire you in a new and profound way.

This is why a key foundation in our relationship with animals is to be fully present. Animals are always living in the moment; it's their instinctual nature. A dog is not counting the bones they chewed over the last month or worried about the squirrel that may enter the property next Friday. The dog is noticing, feeling, and responding to what is happening *in the now.* Because animals are sensitive to energy, they pick up on human energy in any given moment. If we show up carrying the energy of what happened yesterday or last week, this can be confusing for the animal, and it may blur our communication with them. Or, if we show up laser-focused on an end goal, we can't meet them where they are in the moment and will probably not be successful.

When we strengthen our ability to feel the power of the present moment and embrace it, we elevate our awareness and understanding of the world and those we connect with, especially animals. There are always signs, symbols, and messages being shared through nature, but we have to listen to our inner guidance and *decide* to notice! As we take more time to pause and really feel in the moment, listen in the moment, respond in the moment, and truly experience the moment, we build a new level of calm, confident mastery.

In this chapter, we explore foundational ways to attune to the present moment, including journaling, meditation, grounding, and breathwork. All of these practices will support not only your work with animal communication but your well-being.

Journaling

When I was 16 years old, I was feeling called to connect with nature and challenge myself by doing something I had never done before. I asked my parents if I could do Outward Bound, a program that leads groups through outdoor expeditions in remote locations around the world. I chose to go on a canoe expedition in the boundary waters of Canada, 22 days of low-impact camping (which means leaving no trace). My parents were reluctant to have no contact with me but felt my desire to challenge myself and recognized it as a great opportunity for learning and growth. Little did I know what I was getting myself into.

That summer, my mom dropped me off at the airport and I flew to International Falls, Minnesota, to meet my traveling group. This was a time before the Internet and cell phones. Our group of 12 (2 instructors and 10 students) gathered in a van and drove over the Canadian border to a desolate location deep in the wilderness. We were dropped off only with our backpacks, stuffed with clothing, tents, food, and waterproof maps. The experience was more challenging than I realized—mentally, emotionally, and physically. I did not expect to be walking with a twenty-foot canoe balanced on my shoulders every day between lakes, but I sure learned quick.

One of the many gifts of this experience was the opportunity to be in nature. Getting away from city life and all of the typical stresses of high school helped me connect with the energy of the earth. For three straight weeks, this immersion offered a profound way to get in touch with the power of the present moment.

About midway through the expedition, each camper was dropped off at a remote location in the forest. The exercise was to be by ourselves for three days straight with limited rations. One of our trip leaders would circle by to check on us once a day, but there was no talking. We were given a journal at the beginning of the expedition, and it became my outlet for mental and emotional balancing. I wrote in my journal every day, expressing my fears, doubts, and worries as well as my hopes and successes.

During the three days of solitude, I had nothing else to do but sit and be present. At first, I felt mentally and physically uncomfortable. My mind was racing, not used to quiet and stillness. I walked around, pacing, not sure what to do with myself. But after a few hours, I sat down with my journal and surrendered to the moment. I spent the days feeling and absorbing all of nature with my senses. It felt as if my natural surroundings were guiding me to slow down and soak in the experience. I journaled about everything. The more I wrote, the calmer and more balanced I felt, and I began to notice and feel the shifting energies of everything in the world around me. I had never before taken the time to slow down and pay attention in this way, and nature more than delivered.

As I sat alone, I observed many animal species native to the area—moose, deer, beavers, otters, a lynx, a timber wolf across the lake, many birds and bald eagles, and even a black bear. We were warned to be cautious of animals, especially bears, and we were trained in the necessary protocols to take in case we found ourselves in a precarious situation.

On the second morning of solitude, a bear appeared through the thick brush of trees, about seven hundred or so feet away from me. My initial reaction was panic. *I am just a city kid all alone in this bear's habitat—what do I do?* But because I had been journaling and tuning in to the natural environment so deeply, a sense of calm washed over me. I first wrote down in my journal all my fears and worst-case scenarios as a way of releasing my fears and tension. Next, I felt guided to tune in to the bear intuitively and energetically. I experienced an inner feeling that the bear was aware of me and acknowledged me, but he was more focused on gathering food from his usual sources. I could feel his awareness of my fear and that the fear was only making me appear questionable. In response, I quickly shifted my energetic focus to feelings of respect, understanding, and my own calm confidence. I envisioned myself rooted in the ground like an ancient, strong tree, radiating a frequency and bright light of calm and respectful boundaries. Throughout this experience, I was writing everything down. I could feel the bear's reception to the energy I was holding and sharing, and I felt a message from him: "It's all good, just passing through. I see you and feel you; you respect me, and I respect you." The bear moved on and though I did not see him again, the experience was profound. There was a great shift within me, almost like I passed some sort of spiritual test.

I was so excited to tell everyone in my group about my experience, but there was still another day and a half of solitude. It was another challenge and exercise in letting go of the future and continuing to appreciate the present moment. There were many more gifts through interactions with other animals, insects, and even the trees. My journal was there for me to ground my experiences, to not only record them but to explore through words everything I was observing and feeling with all my senses. Interestingly, because the expedition was filled with so many challenges, breakthroughs, and lessons, I forgot many details from those three days of solitude. Years later, I read through that journal again and recognized how deeply that experience set my path for my life's work.

I tell this story to share one of my core practices to help you ground yourself and be present in the moment—journaling. After my Outward Bound expedition, I recognized how helpful it was to have an outlet, and I have had a journaling practice ever since.

Journaling also became a powerful tool that I used—and still use—when communicating and connecting with animals. So many thoughts come and go through our consciousness, and journaling provides an opportunity to capture them raw and in the moment. When I tune in to an animal, I always have a journal handy to write down anything that comes to me. Often, I will go back through what I've written and be surprised, not remembering everything that came through. So it not only helps keep you present, but is a physical reminder of what happened moment to moment.

I encourage you to get a journal to use with this book. The journal will serve as a place to record your thoughts, feelings, observations, and experiences. It will also serve as a tool to ground you in the present moment, helping you to slow down and observe more deeply with all your senses. You may not realize how much energy you have bottled up within you, which may manifest as thinking in circles or feeling uptight, stressed, and emotional. A journal is a great place to release all of that energy, like a mind dump on the page so you are free and clear to be more present. There is something very meditative about journaling because you are quieting the mind and opening up to a flow of consciousness.

Any journal will do: you can use a simple spiral-bound school notebook or find something more sophisticated. Just make sure you do not overthink it and that whatever you pick will be something you'll use. An expensive, leatherbound journal is gorgeous, but if you're afraid you'll spoil it by spelling a word wrong or making messy notes, you'll never use it. I can relate to the feeling of not being able to find the perfect journal, but that only means I end up not journaling at all. Pick one and jump into it.

Practice: Freewriting

Every chapter of this book closes with journal reflections for you to consider as a way of deepening your understanding. But I also recommend making time for freewriting three pages every day.

Freewriting is just what it sounds like: writing freely, without any prompts or aims for a specific result. It is putting pen to paper and writing without stopping until you hit the bottom of that third page. Write down whatever is on your mind, knowing that no one will ever read it. This is only for you. Don't worry about punctuation, your handwriting, or grammar. Just let yourself go. You could write the same sentence over and over, "I don't know what to write. I don't know what to write. . . ." until something else comes out. Maybe you vent about whatever is annoying you, who is bothering you, what keeps you up at night. You may also write about what makes you happy, what you are grateful for, what you love. I like to do these in tandem, spilling out my frustrations and fears and then filling myself back up with what I want to manifest in the future. In the end, it's not about what you write, it's about the process of writing.

If possible, I recommend freewriting in the morning. When we first wake up, our energy is most clear and balanced. We are fresh but still close to the dream state, and many people find it easier to tap into their unconscious selves this way. But in my experience, as long as you write three full notebook pages every day, it does not matter what time of day you do it.

When you practice freewriting regularly, you begin to open up a channel to your inner guidance, releasing blocks related to your awareness of subtle energies and your creativity and allowing your intuition to be fully present and available. Commit to a month of freewriting in your journal, filling three full pages every day, and see what happens.

After a few days of journaling, people often start to experience shifts within their life because they have released thoughts, feelings, and emotions that were bottled up within their energetic system. You may suddenly have an amazing idea to create something or take action in some way. Synchronicities may start to occur; new opportunities may materialize. You may find you are in the right place at the right time, making better choices, and feeling more balanced and fulfilled. It's a powerful practice.

Feeling this way is good for you, and it's good for the animals in your life too. Allowing yourself to release daily will clear your energy so you embody the calm, confident leader your animal needs you to be. This way, when you begin working to communicate with them, you can be fully present to love, connect, and grow to your highest potential.

Meditation

During that same Outward Bound expedition, I also experienced the power of meditation and its benefits deeply for the first time. Our trip leaders didn't call it meditation, though; they simply suggested we sit in the silence among the trees and water, focusing on our breath and enjoying the peacefulness of nature. I took their advice, closing my eyes, inhaling and exhaling deeply, and allowing myself to be present with the beauty all around me. If you've ever struggled to meditate in a specific way, you will be pleased to hear that even this kind of informal practice is meditation!

I found it was very easy to be still and quiet my mind when I was out in the wilderness, away from city lights and the energy of people and technology. Each day we took time to just be with ourselves and sit in the silence. And with each day, my ability to meditate became stronger. Soon it was more than just a calming practice; I could sense the energy of the trees, water, earth, and animals. I was clearing the static and inner chatter of my mind that felt like having multiple television stations playing at once. Through meditation, I was able to turn off all those channels and receive a new singular channel of higher communication and energy that was peaceful, balanced, and fluid.

When I returned home after my three weeks, I experienced culture shock: everything felt so loud and fast. Journaling and meditation became my lifelines, tools that helped me stay centered in that calm, clear energy I'd found in nature. They also gave me a foundation to build upon in my intuitive work and communication with animals. I made it a regular practice to quiet my mind and tune in to my ferret, Rocky, as well as my Irish terrier, Corky. After several minutes of sitting in the silence and breathing with my pets, I would pull out my journal and just see what came to me without any expectation or goal in mind. This process strengthened my ability to sense and interpret subtle energies.

In my experience, consistent meditation practice is the key to effective and clear animal communication. It supports all the other crucial pieces, like improving personal development, energy awareness, and intuition, and strengthening the health of the body, mind, and spirit. When we calm our energy, we can receive our inner guidance. We can also communicate more effectively with animals because our mind is not filled with distracting thoughts that pull us out of the present moment. We are not assessing and evaluating what we experience; we are just *experiencing*. In addition to removing the static in our energetic connections, meditation also helps us stay grounded in our thoughts, feelings, and emotions so we are clear about what we may be picking up from others. A consistent practice of meditation will strengthen your ability to sense the subtle energy in everything and to trust your interpretation of what you receive. You gain an ability to tune out what is not serving you, instead focusing all of your awareness in the direction you desire.

The modern world rewards tangible results, which means we are often caught in a loop of pushing ourselves to always produce something new to stay relevant. This is so tiring! We are also used to being overstimulated, having a busy mind filled with conflicting thoughts and draining emotions. So, it's no wonder that taking a few minutes to be still and silent makes many people antsy. With this mindset, you may spend your meditation session stressing about your list of to-dos and how sitting there is delaying your productivity. But as I've seen with myself and my clients, nothing could be further from the truth. Taking even five minutes a day to meditate will give you more energy to be productive. A clear and focused mind helps you get things done and builds resilience so you can remain calm and balanced in stressful situations.

Ways to Meditate

There are many ways to meditate, and one size does not fit all. If you are someone who has tried seated meditation and continues to struggle with it, there are still plenty of ways you can reap the benefits of meditating. You might try:

- *Body scanning*—a seated or prone meditation where you focus attention on each part of your body in turn, letting go of tension, sensations, and distractions (see page 43).
- *Breathwork*—any practice where you consciously control your inhales and exhales (see page 16).
- *Chanting*—the process of repeating a mantra in a harmonic tone.
- *Guided Meditation*—listening to an audio recording of someone leading you through a specific meditation process. (There are great apps for this, such as Insight Timer, where you can find endless guided meditations and music from teachers all over the world, including me!)
- *Journaling*—especially freewriting—the process of recording personal thoughts, experiences, insights, and emotions in a written format. It often involves keeping a diary or journal where one regularly writes down their reflections on daily events, feelings, and ideas. (See page 4).

- *Mindfulness*—sustained focus on the task or activity at hand. It can be practiced during almost any activity—walking the dog, painting, gardening, washing dishes—as all it requires is sustained focus on the task at hand.
- *Walking Meditation*—the process of taking a walk while practicing breathwork and mindfulness.
- *Yoga*—a form of moving meditation where you concentrate on deep breathing while slowly stretching the body and maintaining poses.

These are simple and effective techniques because they're not about clearing your mind and thinking about nothing. Instead, you focus your attention on one thing—whether it be your body, the blank page, specific words, your breath, or anything else—using that to keep you anchored in the present moment. When we focus on one thing, it calms the busyness in our minds and allows us to release the need to divide our attention, which can be a draining source of stress. This is why the old method of counting sheep helps people fall asleep. When you lie in bed imagining one sheep after the other jumping over a hurdle, your attention on this simple repetition relaxes you.

Though these techniques are all useful, I still recommend developing some kind of seated meditation practice. Perhaps you begin by incorporating more mindfulness into your daily run and then, as you feel clearer and more centered, you give the Grounding and Centering Meditation (see page 14) a go. When beginning any practice of meditation, I recommend setting an intention and committing to it. Choose a kind of meditation that appeals to you and commit to trying it for at least a week. Plan ahead by blocking off that time on your calendar. The best practice, of course, is the one you will *actually* do, so beginning with a short session, maybe five minutes, is a great way of easing into meditation. That said, in my experience, meditating for 10 to 20 minutes a day is most effective to feel the benefits. This is not about getting from point

A to point B or achieving any specific results. Meditation is about the practice itself. Consider it in the same way as brushing your teeth or taking a shower—you can't just brush your teeth once and be set for life. In the same way, make meditation a part of your daily self-care routine.

Still not convinced? Meditating daily does not mean you will be free of stress and the ups and downs of life. But you will be able to respond to daily challenges more quickly and effectively, and you will bounce back to your balanced and centered state with more ease and fewer repercussions.

Here's an example. Recently, I was walking my three Australian shepherds in our neighborhood. We passed a neighbor's house where a German shepherd always barks and lunges along their fence line. We are used to the dog's reaction and continued to walk by unbothered. But this day, the dog pushed through the gate and came charging after us down the street. We turned around to see the extremely angry dog running at us and barking. Within an instant, I made myself large, pulled my dogs behind me, snapped my fingers, and threw energy at the incoming dog with a strong open palm as I said, "STOP!" My dogs all just sat and looked up at me, following my lead. And the German shepherd immediately stopped in its tracks, turned around, and ran back to his caregiver, who had just stepped outside with no idea what was going on.

Because of my regular practice of meditation, I was able to respond in the moment instinctually with clear intention and calm confidence. The incoming angry dog felt my grounded, strong energy and intention, as did my dogs, who gave me room to handle the situation and protect them. A challenging situation doesn't have to be as extreme as an aggressive dog, but with meditation you will be able to move forward without that nasty e-mail, stressful conversation, or parking lot kerfuffle getting the better of you. My dogs and I continued on our outing, and I must admit, I did walk with a confident spring in my step after that. My dogs looked at me all happy and proud, and I said back to them, "Yeah, I did that! I've got you. We're all good."

Practice: Grounding and Centering Meditation

If you feel scattered, imbalanced, foggy, distracted, or even fearful, doubtful, or confused, you could benefit from grounding. When it comes to our animals, being ungrounded sends mixed energetic signals, making it confusing, difficult, or scary for them to connect with us. So, this meditation is a great way to begin your daily meditation practice or at least try a few times each week. When you dive into Part II and start your work with animals, we will begin each session with this grounding meditation. The more you practice, the easier it will be to drop into that centered, relaxed space.

1. Begin by sitting in a chair or comfortably on the ground. Close your eyes and take a deep breath in through your nose, exhaling through your mouth. Settle into your space as you continue to breathe slowly and deeply, feeling your body begin to relax more and more with each breath. Feel the earth below you, always there supporting you, holding you safe and steady.
2. Imagine a grounding cord attached to the base of your spine extending down and connecting to the core of the earth.
3. Imagine a beautiful pillar of golden white light shining down upon you from way up high in the sky. The pillar shines down through the top of your head and shoulders, all the way down through your torso, down your arms, through the palms of your hands, the tips of your fingers, and your legs and feet. This pillar of light shines through you and around you, bathing you in light. Feel the gentle yet powerful light soothe and center you.
4. Take a few minutes to focus on your breath, basking in this centering pillar of light and grounded by the supportive energy of the earth.
5. If you are practicing a deeper meditation or using this as a launching pad for connecting with animals (see Chapter 7), move on to that part of your practice now.

6. When you are ready to release, imagine the pillar of light slowly and gently receding up into the sky. See the grounding cord gently detaching from you and dissolving into the core of the earth—the earth naturally absorbs and recycles the energy. Begin to feel your body in the space that you occupy, gently wiggling your toes, loosening your legs, shifting in your seat, loosening your back, expanding your chest, moving your shoulders, arms, hands, and fingers. Gently rotate your neck and loosen your jaw as you take one more deep breath in and out. When you are ready, open your eyes and come back to your physical space.

Here are a few more tips to consider, whether you're just starting out meditating or if you're looking for new ways to enliven your practice.

- Try listening to soothing meditation music available on meditation apps such as Insight Timer or from playlists on Apple Music, Amazon, Spotify, or YouTube; nature sounds; or any audio that is relaxing. Calm sounds can help to shift your energy and consciousness to a meditative state, especially if you always use the same audio to practice. You'll find you slip into your meditation state more easily as soon as you turn it on.
- Distractions in your environment will likely pop up, whether it's the sound of the refrigerator, a barking dog, a leaf blower outside, or a running fan that you never noticed before. Rather than trying to block out sounds, try tuning in to the sound and using it as your focal point while breathing deeply. What we resist persists. If you surrender to the distraction, allow it, and use it as an opportunity to focus, you diffuse the resistance and frustration while also releasing all of your other thoughts and feelings.

- You may also feel sensations in your body as you start to relax in meditation, such as an itch or an ache. Just notice the feeling and allow it. You may even imagine you are placing that sensation in a bubble and letting it float away as you bring your attention back to the moment.

Breathwork

When we are stuck in our head, overthinking, stressed, or preoccupied, we unconsciously tend to have a shallow breathing pattern. I observe this in clients daily, and it affects our energy and the signals we are projecting, limiting us from truly connecting with our intuition and animals.

Breathing is something so obvious that people often take its profound healing and calming properties for granted. But on some level, we do understand its importance, such as when someone is freaking out and your immediate response to them is "take a deep breath!" We don't have to be panicking for breathwork—or controlled breathing practices—to bring us clarity and improve our relationships with our pets, other people, and ourselves, though.

There have been many times when I have realized that I am not breathing very well. I suddenly take a deep breath and realize, *wow, that's all I needed, I feel so much better.* Sometimes while walking my dogs through busy city streets, they will start pulling on the leash seemingly out of nowhere. *What is happening?* I think to myself. *Why are they behaving this way? They know better!* Then I check in with myself and realize that my breathing is very shallow. I inhale deeply, pause, and release a big, deep exhale—my dogs immediately stop pulling. They are simply feeling my tension. Consciously breathing plugs me back into the depth, clarity, and reality of the present moment.

Whenever I introduce meditation to clients, I always begin the practice with a consistent breathing pattern. In this way, I make breathwork an integral part of meditation. But you don't have to dive deep into meditation to realign yourself. Breathwork is a practice that you can do anytime, anywhere.

Practice: Mindful Breathing

Mindful breathing is any kind of breathing where you are paying attention to and controlling your inhalation and exhalation. Here is an easy technique to drop into your breath.

1. Begin by focusing on your breath. Inhale deeply through your nose for a count of four.
2. Hold your breath at the top for a count of two.
3. Exhale slowly, controlling your breath as it comes out of your mouth for a count of four.

You can feel out what the best speed and pace of counting is for you, but the idea is to go slow, not to hyperventilate. You can focus on this practice for 10 breath cycles, or however long feels right for you. The great thing about this technique is you can do it anywhere and at any time, even with your eyes open, such as while you are training your cat, showering, feeding your child, driving, or even before—and during—a big meeting.

Mindful breathing elevates your energy and vibration to a higher level where you can see and understand everything more clearly. In this state, you also attract other supportive energies and vibrations, which only further boosts the positive effects.

It may seem so obvious, but this process is very healing. As you practice, it will become easier to notice the quality of your breath throughout your day and improve it. You can even set a subtle alarm on your phone to go off every hour, reminding you to practice breathwork—or even just to breathe! The more you find your calm center, the easier it will be for you to understand the needs of your animals and serve them to your highest potential.

Living in the Present

Though we won't get into direct communication techniques until Part II, that doesn't mean you can't start fostering a stronger, more energetically clear bond with your pets right now. For example, when I am out with my dogs, I imagine I am grounding myself within the core of the earth as we walk, paying attention to each sensation: the air on my skin and in my lungs, the leashes in my hands, the flow of our movement together, the beauty of their fur and their stride. I notice what they are paying attention to, imagining how they sense their environment, what they smell, taste, touch, and hear. I pay attention to what I see and hear around me and also what I feel energetically, both as an individual and as a pack appreciating the moment together.

Even when I am just at home having a lovefest with one of my dogs or cats, I am mindful of being present and appreciating our time together, whether it be cuddling, me petting and massaging them, or giving and receiving kisses. I am fully absorbing the joy of the moment together, noticing our connection and celebrating it deep within my heart and through my body and all my cells.

Animals can be our greatest teachers in this practice because they are always living in the moment. When my cat, Peanut, is relaxing in a sunny window, she is present in the experience of basking in the warm rays of light. She is not thinking about TV time on the couch later. Nor is she stewing about the vet appointment last week that she did not enjoy, to say the least. While vet appointments can be quite stressful for her, once they're over and we get home, she steps out of her cat carrier and gets back to her usual routine, responding to whatever that current moment is presenting her.

When we meet animals consciously and energetically where they are, we can transform our mood, receive new ideas and inspiration, and open up a path for clear communication. The animal also benefits as you release conflicting energy because it soothes them and allows them to more fully be themselves, free of distraction. I have a dear memory of being deeply present in a moment of love and cuddling with my dog, Cooper, not long before she passed away at age 15. At the time, I made a conscious and clear intention to be fully in that moment of love with her, absorbing with all my senses, really taking it all in with gratitude and appreciation. To this day, I still vividly remember that moment; I can call upon it as if I am reliving it, feeling the moment and Cooper through all my senses and within my heart. It was a gift to Cooper because I know she appreciated the depth of our understanding and felt my love. And it was a gift to me because I feel healed and blessed by it even today.

Perhaps you're seeing a theme here—all of these preliminary practices are focused on getting you to live right where you are, right now, in the present moment. This is a lifelong practice, and it begins again and again, each moment you decide to simply stop and pay attention. As you move throughout your day—and especially when interacting with your pets—try being truly present in the moment by grounding, breathing, and using all your senses. You and your animals deserve this gift in every moment of every day.

Journaling, meditation, grounding, and mindful breathing will help you create a foundation of effectively feeling, listening, and experiencing what is happening right now. When it comes down to it, isn't that all we really have? In the next chapter, we will build on this by focusing our awareness and setting intentions.

Journal Reflections

1. How do you feel about meditation? What is your resistance to meditation? Allow yourself to write down your doubts as a way of acknowledging them so you can release them.
2. Imagine yourself succeeding at meditating. What would it feel like to be super successful at meditating daily? What could be the benefits? Visualize yourself succeeding and journal about it.

3. What are some other ways to ground and center? Identify other experiences that ground your energy besides the meditation explored in this chapter and set an intention to try them out until you find what works best for you.
4. When you notice your breath and deliberately slow it down, how do you feel? What changes happen mentally, emotionally, and physically? How can you incorporate mindful breathing into your daily habits?

SETTING INTENTIONS AND CULTIVATING AWARENESS

When I was a kid, my dad would regularly take me to dog shows in Chicago, and once, when I was 10, we had the opportunity to meet Barbara Woodhouse, a famous dog trainer and author of the best-selling book *No Bad Dogs: The Woodhouse Way*. She was in her seventies, British, and very proper. I was starstruck and felt inspired by how she improved dog behavior and restored balance to people's homes. When she offered me the opportunity to work a bit with her Scottish terriers at the event, I instantly said yes. She observed my leash handling and interaction with her dogs and said to me, "Kid—you've got *it*! You've got something very special, and one day you are going to be helping many people and their dogs. You remember these words and hold them in your heart!"

I felt as if I were blessed by royalty or a fairy godmother. At that moment, I was seized with a desire to cultivate the talents that Barbara Woodhouse saw in me. She mirrored something I felt within myself that, though I had not yet identified it, I knew to be true. By observing her, I was able to tune in to what it would look and feel like to help people and their pets. The seed was planted, and I began broadcasting thoughts, feelings, and hopes about working with animals to the Universe, even if I did not know exactly how or when that would come about.

Once I settled into my life's work many years later, I was able to look back, connect the dots, and recognize that what I had instinctively done as a kid was create an intention. I can now see how I was always guided along my path and all of the synchronicities that aligned to support me along the way. Even if my life took me in different directions that may have seemed off track, every experience had its purpose and was a building block toward manifesting my intention.

The Power of Intention

People often do not realize how much power they have to create their reality. It is easy to go through life passively reacting to whatever life presents us each day. We respond to our circumstances without any clear idea of what our goals are or how we can use our unique skills to be of service to our highest good and the highest good of others. This may lead to situations we don't want to be in, and leave us wondering, *How did I get here?*

But you can make your dreams become reality, and this begins with *intentions*. There are two main components to setting intentions that will lead to positive and valuable outcomes:

1. *Identify your values and goals*: Having a solid idea about what you desire in every aspect of your life is crucial because, as the saying goes, energy flows where attention goes. The more you visualize yourself living the specifics of your dream, the more you draw that future toward yourself.
2. *Act with intention:* Rather than *reacting* to everything that comes your way in life, clearly defining what you are aiming to achieve empowers you to *act*, to work toward that goal. While of course it can happen that simply naming your intention will help it manifest, this process is easier and faster if you are fully involved, connecting with the feeling of succeeding, planning it out, and tackling the necessary steps.

Our thoughts and feelings create things and experiences. When we have a clear intention and connect to what it would feel like to experience our dreams and desires, it lets the Universe know what we want, and we begin the momentum of creation and manifestation. We transmit a frequency into our world and the Universe, and the Universe responds by providing us with energies and experiences that align with our expectations. The key is then to reach out and grab it. Learning to recognize when the Universe is helping you and responding actively is just as important as setting clear intentions. Otherwise, those opportunities will pass you by.

It's important to remember that intentions work both ways. When our thoughts, feelings, and emotions are focused in a "negative" space, we may create misguided intentions and begin inadvertently attracting energy and experiences that are aligned with those negative vibrations. For example, maybe you argued with someone and find yourself dwelling on it. Later the same day, you may painfully bump your head and then find out you got a parking ticket. You feel like you can't win: "when it rains, it pours!"

When we focus on the negative, it can draw more negative energy to us. The good news is that we can easily break this pattern by shifting our attention to positive thoughts, feelings, and emotions. To be clear: bad things do happen, and it's human nature to experience negative emotions. There is nothing wrong with feeling irritated, apathetic, depressed, or any other range of feelings. The point here is to develop an awareness of where your focus is. If you realize you are dwelling in sadness about not getting that job you applied for or anger at your partner for forgetting to load the dishwasher again, you can course-correct by shifting toward the positive. Perhaps not getting that job means you find another one without such a long commute. Or while you clean the kitchen, you can catch up on your favorite podcast.

When you notice you're focused on the negative, you can see that as one end of a spectrum. Then, you can imagine the opposite of that feeling and pivot your attention. Of course, this is easier said than done. But every moment you recognize you are focusing on the negative is an opportunity to shift to the positive. Visualize, imagine, and feel yourself experience the positive end of the spectrum. Focus on what feels good. Take a few deep breaths. The longer you spend in positive energy, the easier it is to diffuse negativity.

As we move through this book, I invite you to become aware of your intentions for learning and growing. It's so important to get clear on what you would like to create or change. When thinking about your pets, maybe that means consistent communication, solving behavioral issues, discovering new activities to do together, or teaching new tricks or tasks.

Begin to imagine what your life looks like and how you feel as you communicate intuitively with animals with ease and confidence. Even if you have yet to experience any kind of deep or profound animal communication, allow yourself to get creative and imagine it in as much detail as possible. Let go of doubts, fears, or second-guessing, and really tune in to your intentions, for they are the sails you are setting that will catch the winds and steer you toward your desires.

Practice: Intention Journaling

Grab your journal and write down some intentions. Whether related to communicating with animals or not, we all have desires and goals we would like to manifest. Write down, with as much specificity as possible, a vision of the future where you've achieved your goal. What does success look like? How do you feel? How do others feel? Allow your imagination to create the dream scenario of you manifesting what you desire and really connect to the feelings of your success.

Your journal is also a great place to reflect when you have achieved the desired results of your intention setting. The more we are aware of our successes and appreciate them, the easier it is to understand the power of our intentions and feel confident in the process of manifestation.

Understanding Energy and Energetic Vibrations

Quantum physics posits that everything is composed of energy-producing particles that are in constant motion, creating specific vibrations, which I will call *energetic vibrations*. In this way, energy is everywhere and within everything—people, plants, trees, animals, rocks, man-made objects, and even thoughts, feelings, and emotions. The speed at which something vibrates is referred to as its *frequency*.

High frequency vibrations are generally associated with positive qualities and feelings—love, joy, forgiveness, compassion, peace—and have a light, airy radiance. On the other hand, low frequency vibrations feel heavy and thick and are associated with feelings like hate, fear, greed, sickness, and depression. Consider the energetic vibration of a metal telephone poll, which probably feels dense with a slower frequency. In contrast, the energetic vibration of an oak tree may flow freely or even feel pulsating and expressive.

Because of the busy world we live in, people usually are not consciously aware of energy and energetic vibrations. Nor do people typically set an intention to be mindful of their own energy and the energy of everything around them. Our world requires us to stay busy and be productive, and though technology keeps us distracted from subtle frequencies, they are always there. Have you ever entered a room where something did not feel right? Maybe you felt a heavy or sick feeling in your stomach, and you wanted to leave immediately. Or have you ever sat on a beach, watching the waves roll in and out, and just felt a huge weight or fog lift off you? These are examples of our awareness of energy and vibrations.

Recognizing Your Energy

Sometimes we may not realize the energy and energetic vibrations we are bringing to a situation. Often, we are in denial about our feelings because acknowledging how we truly feel may throw us off-balance or reveal weaknesses to ourselves and others. But feeling your feelings is a part of being alive, and the more you embrace and honor your feelings, the more control you have over them.

Let's look at an example. Daniel contacted me because his two cats, one of whom was a more recent addition to his household, were fighting all the time and he had no idea what to do. He wanted me to communicate with them that they needed to get along or one of them would have to be rehomed. As soon as I met Daniel, I could feel his tension. I asked him, "How do you feel right now?" His loud and reactive response was, "I'm fine! I'm good! I feel good! It's them that are not good!" As soon as he heard himself, he paused for a moment and then said, "Or maybe I'm *not* fine?"

He kept telling himself he was calm and collected, but when he took a second to feel his feelings, he realized that wasn't true. After taking a few deep breaths, he acknowledged how frustrated he was with the cat situation and that he had been under a lot of pressure at work. Because of his busy work schedule, he was not able to go to the gym, which was normally a great outlet for him to feel calm and balanced. As soon as he admitted to himself how he truly felt—frustrated, doubtful, and tense—he sensed a huge release. Recognizing where he was energetically helped him see where he needed to be to guide his cats toward getting along.

When I tuned in to his cats, I could feel in them that they were interested in each other, but they needed some time to get to know each other and adjust to living together. They were also picking up on Daniel's tension, and it was not helping their blending process. Daniel created a daily practice of checking in with himself, so he was aware of his energy. When he did not feel good or balanced, he instead focused on how he wanted to feel, redirecting his energy toward feeling better. That way, he cleared the tension he was holding on to, meaning he wouldn't unintentionally share it with his cats. He also started to meditate every day and created new intentions focused on feeling good in every area of his life.

Daniel quickly saw a shift in his cats and their interactions with each other. He felt more patient and committed to taking it day by day, knowing some days would be easier than others as he helped the cats get used to each other. The cats felt less pressure because Daniel was no longer fuming with tense energy. In addition to the positive strides with his pets, Daniel also felt his work life improve. He was less defensive and combative with co-workers and felt more productive. This relief and shift all happened for him because he was more aware of his energetic vibrations and learning how to shift his vibration to higher levels that aligned with joy and harmony.

As we have been discussing, the first step is to slow down and be here now, fully noticing what you are experiencing. Setting an intention to be present activates your consciousness to recognize and respond to energetic vibrations.

Even as I am writing at this very moment, my youngest dog, Luca, is pawing me for attention. I could ignore him because I have to complete this chapter and meet my deadline. But I could also see Luca asking for my attention as an opportunity to pause and be present with him as he lies by my side, sharing some pets and love. I tuned in to the moment and opted to pause with Luca. Taking a quick break to experience the present elevated my vibrational frequency through love, joy, and appreciation for Luca, along with giving me a new burst of energy and inspiration to carry forward in my writing.

Practice: The Four R's

A concept I like to share with clients is the Four R's: Relax, Reflect, Release, and Return. This idea helps you to calm down, ground yourself in the present moment, or tune in to energy. You can do a brief meditation in just a few minutes, though longer sessions can be beneficial too.

- *Relax:* Take a moment to close your eyes and focus on deep, slow inhales and exhales. Clear your mind and settle into the present moment.
- *Reflect:* Reflect on your energy in the present moment—how your body feels and what's on your mind. Whatever comes up, just notice it, recognize it for what it is, and allow yourself to feel how you feel.

- *Release:* Now it's time to let go of whatever is coming up for you that is not serving your highest good. Imagine placing that energy on a cloud or inside a bubble. Then, see and feel it gently floating away into the atmosphere, drifting farther and farther away. If multiple energies are coming up for you to release, repeat this process gently and with ease, continuing to focus on your slow breathing.
- *Return:* When you are ready, bring your attention back to yourself and your body in the present moment. How do you feel? Lighter? More relaxed? Open your eyes and return to your space.

Attuning to the Energy of Others

Not only is it a good practice to be aware of your energy at any given moment, but it's also important to tune in to the energy around you. To illustrate this, let's meet Melinda, a client of mine who wanted to adopt a rescue dog that would be a great playmate for her two young boys, ages six and eight.

When she arrived at the shelter, she walked through the aisles of kennels filled with dogs barking and whining, desperate for new homes and relief. She came across a two-year-old husky named Blue. He was whining and jumping, trying to get her attention. Her heart broke when she saw him. He was so beautiful, and she felt she had to save him. After a few moments of watching him jump all over the place, she thought he would probably be very energetic and playful, and a good match for her boys. She also thought he might have been smiling at her because his mouth was open, tongue hanging out.

But after bringing him home, she felt that she had made a huge mistake. When her two sons would start running around wanting to play, Blue would nip at them, trying to calm them down and create order. Their energy was too much stimulation for him, and on top of that, he had no training or outlets for his energy. Running and jumping around with the kids only made Blue more stressed, and Melinda made the difficult decision to return Blue to the shelter, devastating her kids.

Melinda reached out to me for help because she felt her family needed a dog, but she didn't understand how to choose the right one. Especially when looking to adopt a pet, people usually come into the interaction with a lot on their mind. They are excited and nervous, wondering if this one or that one is the right fit. They are primarily focused on how the animal looks and acts, and maybe the background information that the shelter or adoption location provides. Perhaps you've already guessed what's missing from this scenario—a focus on the *energy* that they are bringing to the moment. But here's the second key piece: they are also ignoring the energy of others around them.

Melinda first began to evaluate her energy, and, with my direction, she practiced tuning in to the energy of her family members as well. This gave her a better understanding of what the energetic flow in her house was like. Then, I pointed out to her that she needed to release her need to "save" a dog and instead tune in to the energy of the potential dog—beyond what she was seeing and hearing—to find the best vibrational match for her family. After several months of working through the techniques I share in this book, she felt ready to go back to the shelter. Melinda soon adopted Lily, a lab mix, and her family couldn't be happier.

Lily was peacefully reclining on a cot in the shelter kennel, wagging her tail when Melinda walked up. Visually, she noticed how Lily did not jump up desperately to greet her, instead wagging her tail joyously from her cot. Melinda tuned in to Lily with all her senses and could feel a very balanced, happy-go-lucky energy. Melinda also appreciated how Lily approached her and sniffed her feet first before looking up with greeting eyes. Then Lily lay down on her back, wanting belly rubs. This was a stark contrast from Blue's anxious jumping, barking, and drooling at their first meeting. When Melinda brought Lily home to her sons, she made sure the introduction was calm and balanced, and she shared my teachings with her family so everyone was on the same page. As they began to run around and play in the yard, Lily galloped along with them, fetching balls they threw and loving it when they hugged her. Lily was the correct vibrational match for Melinda's family.

Now, you may not be looking to adopt a pet right now and might be wondering if Melinda's story applies to you. It does! Remember how everyone and everything has energetic vibrations—from the candle on your nightstand to the robin outside the window to that big rock you like to sit on to that person driving in the next lane over? That means there are endless opportunities to tune in and feel the energies of everything in your world! The following practice offers some specific steps to get started with this. As with all the practices in Part I of this book, it is the *practice* part that is important. While it can take effort at first to tune in, the ability to feel, read, and understand energy and energetic vibrations is like building a muscle. The more we practice, the stronger the muscle becomes. Getting curious about the everyday objects in your house or the nature around your neighborhood not only strengthens your ability to sense energy, but it expands your understanding of the unique and infinite wonder the world provides, which in turn raises your own vibration.

In addition, intentionally tuning in to energetic vibrations is vital for intuitive animal communication. As I've mentioned, animals are keenly aware of energetic vibrations. For example, you may have experienced an animal responding differently to certain people or spaces. I often work with clients who say their dog is great with people, but then a certain someone will pass by or enter their home and their dog is suddenly reactive. Animals certainly have their own kinds of languages, but it is not a language like our conception of one. It's an energetic language; they are always tuning in to the energy around them, evaluating and sensing the inner workings of people and other animals. Through this, they receive communication that helps them better understand and navigate the world as well as information about how to respond to any given situation.

Practice: Sensing Energetic Vibrations

I invite you to begin to recognize the energetic vibrations in everything around you, wherever you go and whatever you do.

1. Begin to let your attention explore with curiosity and allow yourself to feel into your environment. Use all your senses to feel energy in all things.
2. Notice what surrounds you in your space right now and choose something, animate or inanimate, to tune in to and feel its vibration. This can be a powerful experience outside in nature because there is so much life force pumping through trees, plants, minerals, and the earth.
3. Close your eyes and see what you pick up. For me right now, there is an orchid on a table across the room. As I close my eyes, I tune in to the orchid and feel the moisture it's absorbing from the water I gave it earlier. I feel a coolness and a damp texture at the base. I also imagine nourishment flowing up into the green leaves. I can feel the leaves slowly and gently expanding. I can also feel the tiny flower buds beginning to sprout. The overcast light from outside is gently kissing the surface of the leaves, adding a touch of warmth.

You don't have to spend a lot of time on any one thing. But as you begin to scan your environment with curiosity and feel into one thing after another, you will notice how it gets easier to tune in and receive impressions quickly. You can tune in to anything—your desk, a rock, a car, a person, an animal. The idea here is to get comfortable tuning in without an agenda: the goal is to be present, experience, and to strengthen that muscle.

While your eyes don't have to be closed for this kind of practice to work, I encourage you to do so if it is available to you because it opens up your other senses. When we practice with our eyes closed, we cultivate a stronger ability to respond quickly to whatever is happening, making choices that will better serve us and others. This is because we are acting on intuition, simply knowing and responding accordingly without engaging the logical mind.

Several years ago, I was walking my dogs, Dasher and Harlow, in my neighborhood on our usual route. I began to feel strong, buzzing vibrations within my body, and I sensed I was picking this up from somewhere in my environment. There was a warm feeling about it and a sensation of something on the verge of exploding. I scanned the area to see if anything aligned with what I sensed, but nothing matched up. It was very quiet and peaceful, yet the buzzing warmth remained. We approached a corner among several old palm trees that my dogs loved to pass (there were always a few local cats hanging out there). Instead of walking through the palm trees though, I decided to cross the street, without really knowing why. Suddenly, a gigantic beehive came crashing down out of the palms and exploded on the sidewalk. What looked like thousands of bees were swarming everywhere—right where we always walked. If we had continued on that path, that hive very well could have fallen right on top of us. Fortunately, no one was stung.

Because I was tuned in to my surroundings, my intuition picked up on the energy and guided me to move. My response was very instinctual; I couldn't rationalize anything since it all happened so fast. Because I have practiced strengthening my intuitive and energetic muscle, I was able to sense my environment and respond accordingly, plus my dogs followed my lead. They could not have listened to me or panicked and gone into flight mode. But they remained calm because I was calm.

While tuning in to living beings and inanimate objects will be one of the most useful skills you can develop, it is only the beginning. The following exercise strengthens your abilities to instinctually tap into your environment and feel beyond your primary senses. It moves you out of your mind, which intellectualizes and judges everything, and brings you into a state where you can feel the vibrations of things in their purest form.

This exercise also enhances our ability to understand and communicate with our animals because our awareness of what is happening around us—beyond what we can visually, audibly, or tangibly perceive—is increasing. This means we will be able to feel the energy of our animals and the frequencies they are broadcasting. We will feel what our animal wants, needs, and even what they are about to do or not do. Soon you will start feeling like you have enhanced your psychic awareness and understanding of energetic frequencies, both subtle and profound.

Practice: Energy Exploration

To expand on the Sensing Energetic Vibrations practice on page 31, you can strengthen your ability by practicing sensing beyond your environment.

1. Begin by sitting or standing somewhere in your home. Close your eyes and imagine what is in front of you. What do you feel? For example, maybe there is a glass of water on a table. With your eyes closed, notice all the details in the glass: colors, shape, texture, reflections, ripples in the water. Feel into every detail. Then notice other things around you and all the details that make them up.
2. As you scan your immediate environment, extend your awareness in front of you, into another room or whatever is beyond your personal space. Notice the details of everything and also pay attention to what you feel.
3. Next, extend your awareness even farther, maybe out into the street or backyard, and explore what is there in detail, feeling everything.
4. You can continue to extend your awareness farther out into your town or city, noticing and feeling, and then even farther out beyond your state lines, country borders, coastal lines, all the way out into the Universe, noticing and feeling, only using your senses and imagination.

It's up to you how you would like to practice this exercise. You can practice it in all directions—front and back, to the left and right, above and below—sensing, feeling, exploring, noticing, all with your eyes closed.

Creating clear intentions means being mindful of where your attention is and what you are feeling in the present moment. Practice feeling what you would like to create as if you are experiencing it right now and take inspired action to bring your desires to life. Working with intentions helps you connect with and shift the energy and energetic vibrations you are putting out into the world, and in doing so, you begin to change your experience and draw into your life the feelings and experiences you set out to create. In the next chapter, we dive deeper into our awareness and understanding of how our mind, heart, spirit, and intuition weave together to light our path and help us in our animal communication.

Journal Reflections

1. What does *intention* mean to you? How do you cultivate intention? What does it look like to move through life intentionally?
2. Think about what your biggest dreams and goals were when you were young. How have your intentions changed throughout your life? How are they the same?
3. How would you describe your energy? Use all your senses—what does it feel, look, taste, smell, and sound like?
4. Think of a time when you felt affected by the energy of a place, space, person, or animal. How did that energy make you feel? How did you respond? How are the energetic vibrations of various things or beings similar and different?

THE MIND, HEART, SPIRIT, INTUITION CONNECTION

In Cooper's fifteenth year, I suddenly felt guided to adopt another dog. I initially kept thinking that adding a dog to the house would not be the best idea because Cooper was older and picky about the company of others. But my intuition was telling me otherwise. I checked in with Cooper, intuitively asking her how she felt about it. And the response I felt from her was slight interest and curiosity, combined with a feeling of surrender to the option, almost like, "whatever . . . someone new, interesting . . . I have no issues." My feeling was that it could be good for her to have someone younger around to lift her spirits and motivate her to keep active. I listened to my intuition and explored listings online about available dogs. I visited local rescue events and every city shelter in the greater Los Angeles area. I met countless dogs, and my search became all I could think about, as if I were a magnet being pulled to another magnet. I could feel when a dog was not that magnetic match.

Best Friends Animal Society had a rescue event in the parking lot of a Petco that I swung by, and I met a dog that felt great, but I was unsure. I went home and later that afternoon, I felt guided to go to the Best Friends headquarters to see that dog again. I had no idea where they were located so I looked them up and saw that they were a half hour away. It was 5 P.M., and they closed at six. My mind was telling me it was too late to drive over there, that I should just wait until they opened again. But my gut was telling me to jump in the car and go now! So I headed over. I roamed around their big, beautiful facility, filled with large kennels and tons of rescued dogs. I was not seeing the dog I met earlier that day, so I stopped a volunteer and asked where he was. She told me that the dogs from the adoption events had not returned yet.

But when I told her about my dog, Cooper, and that I was looking for a companion for her, she felt there was a dog I should meet. They had a two-year-old, red-nosed pit bull mix that had just been pulled from the South LA Animal Shelter and was not on the adoption floor yet because they still needed to have a veterinarian examine her. As soon as I saw her, I felt something shift within me, and my spirits lifted. She was very calm, lying on the ground, and she gently got up to approach me with a twinkle in her light brown eyes. As her tail wagged, she laid down on her back so I could rub her belly with such tenderness. She was the one, I could feel it in my soul, and I understood why I was guided to drive out to Best Friends so late that day.

My intuition told me there was information to discover about this dog's story. I googled her name, breed, and the shelter and found a Facebook page where hundreds of people had been following her story and were raising money to rescue her from the shelter. Apparently, she was born at the South Central shelter and adopted from there. The guardian returned her when she was two years old because "she escapes." She had been there for two months and was on death row. The day before she was going to be euthanized, Best Friends pulled her from the shelter.

When they opened back up, I returned with Cooper so we could do a proper meet and greet. The two of them got along really well! Cooper was indifferent at first, and the new dog gently licked Cooper's face, as a way of essentially "kissing the ring" and showing respect, which quickly put a spring in her step. Right away I could feel this was a great match. I signed the adoption papers and named her Harlow.

I share this story to show how we engage with the world through our mind, heart, spirit, and intuition. Each aspect serves us differently as we navigate through life, but sometimes, some of them may hold us back from moving forward with ease and clarity. For example, had I listened to my mind—my rational thought that said it was too late to go to the shelter that day—I might never have met Harlow. Instead, I followed my intuition, even without knowing exactly why. Then I made it a priority to engage with Harlow's spirit, and *listened* when my heart said she was the one.

Having an awareness of how we interact with the world through each of these four aspects of self gives us a clearer sense of what is and is not serving us, so we can make better choices and create balance for ourselves. Let's take a look at mind, heart, spirit, and intuition and how we can begin consciously listening to them.

Mind

As humans, we live through, and in, our minds in various ways. The mind is the center of our consciousness, thoughts, understanding, feelings, beliefs, and experiences. It is the source of our perception, reasoning, and decision-making abilities. Though the senses—sight, smell, touch, hearing, and taste—are what take in stimuli from the world around us, it is the mind that allows us to perceive and interpret those stimuli. It enables us to think, analyze, and process information and helps us to form opinions, make judgments, and solve problems. In addition, the mind stores and helps us connect to our memories, emotions, imagination, and creativity.

Monkeys and Gremlins

Clearly, the mind is what guides us in survival. It may even feel like it is always in charge, at the steering wheel of our existence. But sometimes the mind can get the better of us, such as when we are overthinking or obsessing. This can get in our way and hold us back, distracting us with stories and assumed conclusions that are not reality.

If you have been practicing being in the present moment, perhaps you have noticed that even despite your best efforts, your mind is constantly speculating about the future, ruminating on the past, and replaying old stories. Buddhists call this constant internal chatter *monkey mind.* When a person is experiencing monkey mind, it means that their thoughts are racing, jumping from one idea to the next, similar to how a monkey jumps from branch to branch in a tree. While the mind's nimble machinations allow us to learn from our choices and their outcomes, a choice we made in the past may not apply to who we are today in the present moment. In addition, being bombarded by too many thoughts can create stress, distraction, and overwhelm, hampering us from making good decisions—or any decisions at all.

The key to calming your monkey mind is noticing that it is happening. Then, you can take a break to breathe or do a quick meditation like the following. A daily practice of meditation trains you to recognize and deal with monkey mind because while meditating, your focus is on remaining in the present moment.

Practice: Slowing the Clock Meditation

This method helps calm monkey mind whenever you feel yourself thinking too much or feeling stressed or anxious. Focusing on something specific while following a slow breathing pattern diffuses all the extra thoughts that are distracting and not serving you.

1. Find a quiet and comfortable place to sit or lie down where you can relax and focus without distraction for about 10 minutes. Close your eyes and take a few slow, deep breaths.
2. Imagine a clock in front of you and hear the constant ticktock sound of the clock's second hand.

3. As you slowly inhale through your nose and exhale through your mouth with each breath, keep your attention on the second hand moving clockwise around the clock two times.
4. When the second hand hits the twelve o'clock position again, begin to see, hear, and feel the second hand moving at half speed. Hear the clock going from "tick, tock, tick, tock" to "tick . . . tock . . . tick . . . tock."
5. Continue to follow the second hand around the clock three times at this slower pace.
6. After the three circulations around at this pace, slow down the pace of the second hand even more. Follow the clock for three more rotations.
7. When you are ready, gently open your eyes and awareness and come back to your space.

As our mind operates in overdrive, this also often ignites our *inner saboteur*, a term used to describe the self-destructive thoughts, feelings, and patterns of behavior that undermine our goals and intentions for success. Our negative experiences throughout life cultivate the inner saboteur, but its responses are based on *past* experiences. Though of course it is important to learn to not touch that hot stove, if we turn that into a fear of touching anything red, we are severely limiting our experience of the world. What we resist persists.

Using your imagination to give a face and personality to your inner saboteur will help to address it—I like to call it a gremlin. This gremlin can pop up when you are considering trying something new, maybe taking a new class, considering a new career path, or expressing your individuality in a new way. The gremlin is that voice inside saying, "Don't do it—what if it doesn't go well? You're going to fail, and everyone will laugh at you!" The gremlin was born sometime in the past when you failed or felt hurt. They have assumed the job of taking care of you and letting you know when it is not safe to move forward and make a bold decision.

Their intentions are good—they want to keep you safe, and if you expand beyond what is familiar and comfortable, you may get hurt. But their opinions are based on what has happened to you in the past, not who you are today.

Let's take a look at Ben and Starlight's story to illustrate this. Ben always adored horses and wanted one of his own. When he was nine years old, he went horseback riding for the first time and loved it—until the horse was spooked by a backfiring car. Ben was thrown from the horse and broke his arm. He recovered but never got over the experience. He was so excited to finally ride a horse, and then it ended badly. He never felt comfortable going near a horse after that, but he still loved horses, and every day the accident was on his mind. Even though, logically, he knew it was the loud noise that startled the horse, Ben often wondered what he did wrong to cause the horse to throw him. He also fantasized about if things had gone right, and he had lived a life filled with horseback riding. His mind—his monkey mind!—obsessed over the experience up until his thirtieth birthday, when he felt guided to adopt a horse named Starlight as a way of healing his past.

But he could not shake the story of what happened to him as a child, and there was a little voice in his head that kept telling him that having a horse, much less riding one, was a terrible idea. He could get hurt, and what if he hurt the horse? He couldn't shake this gremlin, and the mental anguish was consuming him. To make matters worse, Starlight would back away whenever Ben approached him. Ben came to me for help. When I connected with Starlight, I could feel that he was willing to allow Ben to ride him, but Ben's energy was too intense and out of balance. Starlight did not want that energy in his personal space. I shared with Ben that he needed to let go of the past and old stories. It was a new day and a new moment; if he wanted to develop a relationship with Starlight, he needed to clear his mind and embrace who he was that day.

I shared with Ben the following meditation to calm the gremlin so he could free himself from the past and embrace the secure, confident person he was. Ben created a daily habit of meditating to calm his mind and clear his limiting beliefs. He stuck with it and just one month later, he was able to get on Starlight and ride! Not only that, but through Ben's meditation practice, he started to feel an intuitive and energetic connection with Starlight. This was a huge breakthrough for both of them, and they literally rode off happily into the sunset.

So, it is important to have an objective awareness of when we are too caught up in our mind, when the old stories are distracting us from experiencing all that the present moment has to offer. When we shine a light on the shadows in our psyche by facing and acknowledging our limiting beliefs, we begin the process of releasing them. This creates space for us to be open to all the new experiences that await us.

Practice: Calm the Inner Voice

This is structured as an informal meditative practice, though it is also something you could do as part of a longer seated meditation. You can try this out on an as-needed basis: when you realize that your monkey mind is spinning and your gremlins are taking control, take some time to unpack what's going on so you can release your fear instead of pushing it down.

1. Find a quiet and comfortable place to sit or lie down, where you can relax and focus without distraction for about 10 minutes. Close your eyes and take a few slow deep breaths.
2. Imagine you are seated at a table. Tune in to your gremlin, that inner voice that is swaying you away from making a decision. What do they look like? Allow your imagination to see and feel them however they present.
3. Invite them to sit at the table opposite you. Allow them to share their concerns and all the potential outcomes that will not work in your favor. Listen, and then share that you are not the person you were in the past, and that their guidance and suggestions are not serving the highest good of who you are today. Express your love and appreciation for them. You may even imagine you are holding them with gratitude.

4. Give them a new assignment, a new job that will support who you are today. For example, rather than them sounding the alarm, you might invite them to be your runner, finding you new opportunities that will fill your life with joy and success. They just need to be redirected and given a new responsibility on your behalf. Thank them and send them off on their new task.
5. When you are ready, gently open your eyes and awareness and come back to your space.

Animals and the Mind

While animals have brains and nervous systems that allow them to sense, process, and respond to their environment, they have a different level of consciousness and self-awareness than humans and do not live through the mind in the same way.

Animals' behaviors and actions are largely driven by instinct and learned behaviors rather than conscious decision-making. For example, a squirrel builds a nest and gathers food for its young because of instinctual behavior inherited through generations of evolution, not because they are actively deciding to do it. In addition, while some animals display some level of cognitive abilities, such as problem-solving (an octopus unscrewing a jar) and communication (a cat meowing by the door to be let out), these abilities are still on a different level from human intelligence and consciousness. Humans can think abstractly, beyond what they can perceive with their senses; think of concepts like mathematics, philosophy, and art. Most animals are not capable of such abstract thinking. Humans also have a different sense of self-awareness, meaning they are aware of their existence and can reflect on their thoughts and emotions. Animals, on the other hand, are aware of themselves in that they may recognize how their presence and energy affect others in social interactions, but they do not analyze or judge themselves.

Living in the mind as humans do can be draining. It can distract us from utilizing our other, more intuitive senses, which in turn affects our communication with animals. The good news is that even if we aren't actively paying attention, we are always absorbing information from our environment. So, the more we get out of our head and let go of the old stories and assumptions, the easier it will be to tune in to our intuition and the subtle energies. This is the plane that animals live in, so becoming attuned to it means easier connection with animals. We begin to meet animals where *they are* energetically, out of the mind and aligned in the present moment.

Practice: Body Scan Meditation

A body scan is when you take a quiet moment to mentally inspect parts of your body, connecting into how they feel individually. This practice gets you out of your head and into your body, which can only exist in the present moment, and it is often used to relax tension that you may not be aware you are holding. If you have trouble falling asleep, a body scan practice while lying in bed can be extremely helpful, though it can also be used throughout the day or as part of a longer meditation. Here's a simple technique to get you started.

1. Find a quiet and comfortable place to sit or lie down, where you can relax and focus without distraction for about 10 minutes. Close your eyes and take a few deep breaths.

2. Now you will travel slowly through your body from top to bottom, moving your awareness from part to part. You can use the following list for reference. At each part, notice any sensations you may be feeling in this area, such as warmth, tingling, tension, or discomfort. Observe these sensations without judgment or analysis and then release them while maintaining a slow, smooth breathing pattern.

- Top of head
- Forehead
- Eyebrows
- Eyes
- Cheeks
- Jaw
- Neck
- Shoulders
- Arms
- Hands
- Chest
- Stomach
- Lower back
- Hips
- Thighs
- Knees
- Calves
- Feet

- Take a few deep breaths and notice how your body feels as a whole. Allow yourself to relax and release any remaining tension or stress.
- When you are ready, gently open your eyes and awareness and come back to your space.

Heart

While the mind is the center of our intellect, the heart is the seat of our emotions. More than just a physical organ, the heart is a powerful energy center that influences our thoughts, feelings, and behavior on a deep level. It is through the heart that we experience love, compassion, joy, and other positive emotions associated with spiritual growth and awakening. In some spiritual traditions, such as Hinduism, Buddhism, and Sufism, the heart is also the gateway to our higher consciousness and spiritual connection. In addition, many spiritual practices focus on opening and activating the heart chakra, which is one of the seven energy centers in the body. The heart chakra is associated with love, compassion, and emotional balance, and is believed to be an important gateway to spiritual growth and enlightenment. By cultivating an open heart, we can tap into the energy of the Universe or divine consciousness that connects all living beings.

Emotions are an essential part of the human experience and play a crucial role in how we interpret and respond to the world around us. At their core, emotions influence our behavior by providing us with information so we can make decisions. If, for instance, you feel happy and satisfied after eating a certain food, you're more likely to seek out that food in the future. Emotions are also an important part of social communication, allowing us to express our feelings and connect with others to build relationships.

Whether we realize it or not, we're always responding to the world emotionally. News stories, work pressures, family dramas, surprise windfalls, health issues, and more take us on an internal roller coaster. Why is this? Similar to the way that the inner saboteur works, our emotions are meant to help us gauge the safety of a situation so we can innately respond to survive. They influence our perception by making things appear more or less important, appealing, or threatening than they might otherwise seem. For example, someone who is feeling anxious may perceive a situation as more dangerous or challenging than someone who is feeling relaxed. In that heightened state, the person is primed for a negative consequence, which might lead them to be more cautious. Our emotions can even influence our physical sensations, and vice versa. For example, if you are feeling stressed, you may experience physical symptoms such as headaches, muscle tension, or digestive issues. Similarly, if you are experiencing physical pain, it may impact your mood and emotional well-being, a trigger meant to help you realize that something is wrong so you can get help.

While this was certainly an important evolutionary tool, it can work against us. We are emotionally triggered by what we see and experience, and our mind often takes whatever ball the emotions pass and runs with it. This can lead to getting lost in our emotions, leaving us exhausted and out of balance, unable to respond appropriately to what's going on. Even just a slight disagreement during a conversation immediately sends alert signals to the brain, which instigates heightened emotions and perhaps a racing heart or perspiration. Reacting emotionally may lead to regret and overthinking, further feeding our shame, frustration, and guilt. People who have had trauma or are experiencing burnout may also become unemotional or emotionally detached, their body and consciousness blocking their emotional response as a protection mechanism.

All that's to say: emotions carry energy. From an energetic perspective, when we are experiencing high levels of emotions, we emit an energy that animals, and sometimes other people, can feel. My mother tells me that even if I am putting on a happy face, she can feel when I am seething with annoyance. On the flipside, when we feel excited, others feel that too and may say, "I love your energy! I just want to be around you because you make me feel good!"

The most important thing to know about emotions is that they are adaptable, and you can train yourself to notice and regulate them so they don't take the reins of your daily life. This is another great reason to be continually working on the practices from the first few chapters. The more you tune in to the energetics of your own emotions and settle into calm, the more you will be able to tune in to the emotions, thoughts, and desires of other people and animals.

The Power of Love

We communicate to animals through our emotions, whether we are aware of it or not. When these emotional communications contradict what we are asking of the animal, they become confused. For example, a dog might be freaking out due to thunder or fireworks. If we tell them it's okay and not to be frightened, but we are scared for them and triggered by their fear, the dog feels those negative emotions. We are saying words to calm them down, but our emotions are conveying anxiety, which confuses them and exacerbates their fear.

Effective animal communication requires awareness of what emotions we are sharing in the present moment. As humans, we are going to have some bad days and emotional moments, and we are allowed to experience those without worrying that we are traumatizing our animals. But emotions, like thoughts, are not always tied to what is currently happening. When we are *aware* of our emotions, we can take responsibility for them and set the intention to share more supportive emotions with our animals.

The highest energetic vibration and emotion is love, housed within our hearts. Focusing on love, feeling it, and imagining we are communicating by sending love through waves, light, or even color is an effective way of supporting our animals.

A simple method to help get you out of an emotional funk where you are focused on negative feelings is by connecting into love. *Loving-kindness* is a type of meditation practice that focuses on cultivating feelings of love and compassion for oneself and others, to reduce negative emotions such as anger and resentment and promote positive emotions such as joy and gratitude. It is a traditional Buddhist practice that has gained popularity in the West as a way to promote emotional well-being and reduce stress. Research has suggested that loving-kindness meditation has a range of benefits, including reducing stress, improving emotional regulation, increasing feelings of social connection, and promoting overall well-being.

Practice: Loving-Kindness Meditation

You can try this meditation when you are feeling overwhelmed by stress or negative emotions. There are many ways to practice loving-kindness, so be sure to seek out other versions or adapt this however works best for you. It can be practiced alone or as part of a longer meditation practice.

1. Find a quiet and comfortable place to sit or lie down, where you can relax and focus without distraction for about 10 minutes. Close your eyes, and inhale and exhale for a few deep breaths.
2. Bring your awareness to your heart center, where love resides within you, and imagine your heart beginning to fill with love and light.
3. Envision the love and light within your heart center becoming brighter, shining like a sun. See and feel it filling your whole body. Your being begins to radiate light and love from every cell within you, shining in all directions around, above, and below you.

4. Through this radiant light, focus your attention on sending positive feelings of love, kindness, and compassion to yourself and others. You may even see and feel this energy of love blanketing the entire planet and extending throughout the Universe.
5. When you are ready, gently open your eyes and come back to your space.

Tapping

People sometimes do not realize how much of their emotions are bottled up within them and draining them. We often bury our negative emotions to cope because we feel that if we face them, they may get the better of us. But when we take the opportunity to acknowledge those emotions, bring them out of the shadows, and give them a voice, we release them and restore balance within ourselves.

The Emotional Freedom Technique (EFT), also known as *tapping*, is a therapeutic method that combines elements of traditional Chinese acupressure and modern psychology to help alleviate emotional and physical distress. The technique involves tapping on specific points on the body, known as meridian points or energy centers, while focusing on thoughts or feelings that are weighing you down and no longer serving you. The idea behind EFT is to release blockages and balance the body's energy system, which can help reduce negative emotions, such as anxiety, fear, and stress, and promote healing. EFT has been used to address a wide range of issues, including depression, phobias, trauma, physical pain, and more. I often recommend tapping to my clients as a way of releasing their anger or fears that are affecting their communication and relationships with their animals. Tapping can be done on your own or with the guidance of a practitioner. To learn more about tapping, I recommend looking into the work of Nick Ortner.

Here's an example of how EFT can help. Linda operated a dog rescue out of her home, often caring for as many as twenty dogs that she pulled from high-kill animal shelters. Seeing abandoned dogs pulled at her heartstrings, and she wanted to give them the love and life that was taken away from them. Many of the dogs had health issues, and living alone, she was the sole person responsible for attending to all their needs while also trying to find homes for them. Linda was led by her emotions to save as many animals as possible. But this was draining her and affecting her relationships with the animals in her care. Though she had always felt she had an innate gift for communicating with animals, she was constantly dealing with misbehavior and chaotic interactions. She felt overwhelmed, defeated, and then guilty for her frustration with the situation she created. She recognized that feeling out of balance was affecting her communication ability and reached out to me for help. I recommended that Linda practice tapping because she was carrying so much stress rooted in emotions—she was chronically in a state of fight-or-flight.

Fight-or-flight is a physiological response that occurs in reaction to a perceived threat or danger. It is a natural and automatic response that prepares the body to either confront the threat or flee from it. When the fight-or-flight response is triggered, the body releases hormones such as adrenaline and cortisol, which increase heart rate, blood pressure, and breathing rate, among other things, so that we can respond quickly and effectively to dangerous situations. While fight-or-flight can be helpful in short-term situations, chronic activation of this response has negative effects on our health and well-being. The process of tapping, though, has been shown to lower cortisol levels and restore homeostasis so we can process our emotions with ease and respond to everyday situations rationally.

I helped Linda through the following tapping practice, and after several minutes, she felt a huge shift in her energy. She took a deep breath as if a weight was lifted, and tears of joy began to stream down her face. After the session, she saw an immediate shift in the energy of the dogs; they were calmer and listened to her guidance. She was able to communicate with them free of all the confusing emotions. I encouraged Linda to take more time for herself and to commit to a practice of self-care. Linda designated 20 minutes a day to tap on her negative thoughts, feelings, and emotions over the next few weeks, and she realigned with her intentions for doing rescue work. She felt more confident in herself too, which the dogs could sense. Restoring balance through tapping gave her clarity to make better choices as she managed the dogs. And this restored emotional balance also helped her to identify wonderful, stable forever homes for them.

Practice: Tapping

With tapping, one session can be very transformative. If you do not feel anything right away, though, don't worry! That's completely fine. Tapping creates deep shifts, recalibrating the nervous system and rewiring neurotransmitters. Sometimes the effects are immediate, and sometimes you won't feel anything until you wake up the next day. It depends on the person and the emotions being released.

1. Sit down with your journal and consider what thoughts, feelings, and emotions you would like to release. Maybe you recognize there are emotions your animals are responding to, such as tension, worry, or guilt. Whatever the emotions might be, spend a few minutes writing down your feelings, even the ones you may be afraid to admit to yourself. Be honest. Remember, this is only for your eyes, and it's an opportunity to acknowledge your feelings so you can release them and restore balance. This is your tapping "script."
2. Once you have your script, find a quiet and comfortable place to sit or lie down, where you can relax and focus without distraction for about 10 minutes. Close your eyes and take a few deep breaths.
3. Begin to tap on the acupressure points on your body (refer to the diagram) while giving a voice to what you wrote down.

Typically, tapping begins on the outer side of whichever hand you choose, and you make three statements there before moving on to the top of the head. You can tap on one side of the body or both sides at the same time, cycling through this list of points:

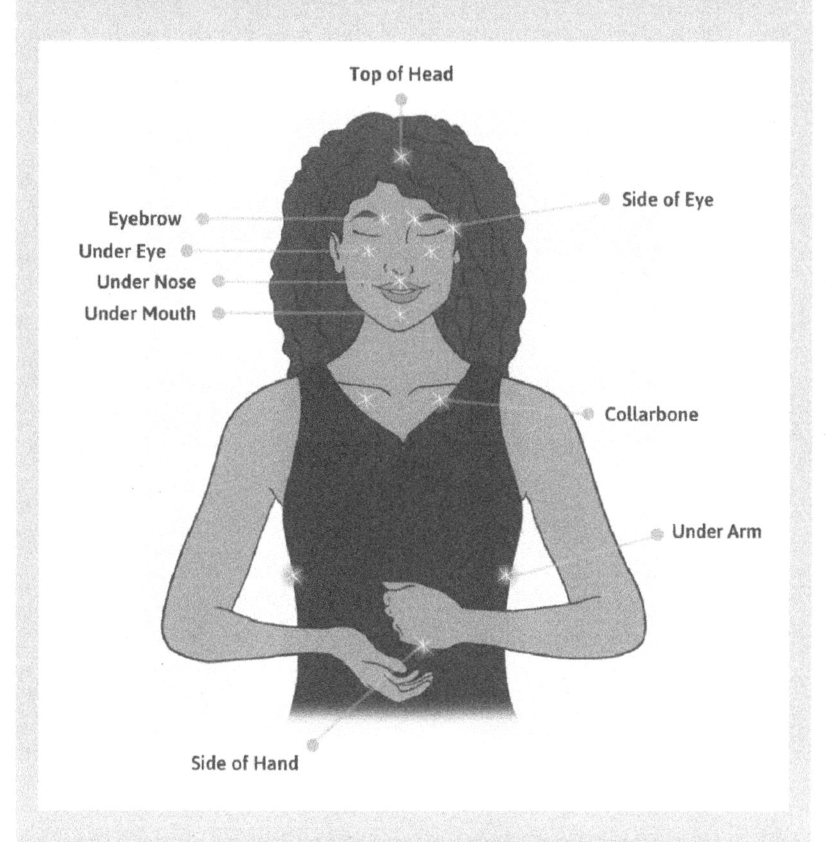

Outer side of the hand (also known as the "karate chop" point)

- The center, top of the head
- Inner eyebrow
- Outer eyebrow
- Under the eye

  - Under the nose
  - Under the mouth (the upper ridge of the chin)
  - The collar bone (where the chest meets the collar bone)
  - Under the arm (the back muscle or bra line)

- With each tapping point, say a statement out loud and feel what you are saying. Then move on to the next tapping point and tap as you say another statement. If you feel yourself starting to cry or even wanting to shout, go with it. This is part of the process. For Linda, this sounded like: "I am tense, I am frustrated, and I am overwhelmed. I have too many dogs! What did I get myself into? What if I can't find homes for these dogs? I feel so guilty. I can't do this all by myself! I am so mad at myself for taking so much on! I am so angry at the shelters that euthanize animals that deserve good homes! These poor animals! I look at them and feel sadness about this world. Why them? They don't deserve this life. And I feel horrible for not feeling stable right now. They deserve better." Continue to cycle through all the tapping points and express your negative feelings.
- After you feel complete, having expressed the negative feelings, take a deep breath, then shift to reciting positive statements. To get in the flow, you could write some positive statements in your journal first. Or you could just start tapping while reciting the positive statements out loud. For Linda, this sounded like: "Even though I've been carrying all of this stress and tension, I love and accept myself. Even though I feel I am in over my head, I am doing the best I can, and I honor my good intentions. Even though I am so angry at the world for the treatment of animals, I see progress, and I am a part of the solution making a difference. I am doing the best I can. I am in the process of taking better care of myself so I can care for the dogs. I have helped so many wonderful dogs and I have a gift for finding them amazing homes." Keep tapping and say out loud a positive statement as you connect into the good feeling of what you are expressing. Continue through the tapping points, focusing on the positive until you feel complete.

- You may feel a rush of joy or the urge to take a deep breath. You may even feel a smile come to your face. Recognize the shifts in your mind, body, and spirit with awareness and appreciation. Allow yourself a few moments to take a few more deep breaths and maybe even shake out your body as a way to circulate the new elevated energetic vibrations.

Spirit

Generally, *spirit* refers to the essence or nature of something, such as the spirit of a person, group, or culture. Maybe you know someone full of spirit—life-force energy pulsates through them that you can feel or even see. Animals freely and easily express their spirit—imagine a dog running enthusiastically for a frisbee or a pod of dolphins leaping out of the water together, clicking, whistling, and slapping their fins. If you have ever felt uplifted in the presence of an animal, you are experiencing their spirit.

In various cultures and religions, Spirit also refers to an immaterial, nonphysical entity or essence that is believed to animate living things or be a manifestation of the divine. It could be a human, animal, or supernatural being, but also applies to plants and other natural things. For example, the spirit of a flower bouquet can be felt through its beauty and unique combination of colors, textures, and scents, as well as its life cycle of blooming and withering. Spirit can refer to a nonphysical essence or force that is often associated with the idea of universal consciousness—the divine spark, life force, or soul—that exists beyond the material realm, transcends the limitations of time and space, and connects all beings and things in the Universe. This metaphysical concept of Spirit emphasizes the interconnectedness and interdependence of all things and the belief that there is a higher, spiritual dimension that transcends the physical world.

When it comes to our animals, some people feel a deep spiritual connection, even identifying a pet as a soulmate they have known for many lifetimes. Other people may feel that when they are with animals, they are connected to something beyond themselves, such as a sense of higher meaning. No matter what you feel in relation to your connection with your animals or your connection with Spirit, one thing is true: having this experience can bring meaning into your life.

Feeding Your Spirit

Our spirit helps us cultivate a sense of purpose and fulfillment, enhancing our overall quality of life. To *feed your spirit* means to engage in activities or practices that nurture your inner self, promote personal growth, and enhance your overall sense of well-being. It involves taking care of your emotional, mental, and spiritual needs in addition to your physical needs. Over time, if your spirit is not fed, you may find yourself dealing with imbalances such as illness or getting stuck in negative emotions. This same goes for animals—if an animal's spirit is not fed through healthy outlets, the resulting imbalance may cause behavior challenges or health issues.

People often forget to feed their spirit because our world is busy and demanding. There are places to be, responsibilities to meet. Then, we wonder why we are so stressed out, short-tempered, or exhausted! When we don't feed our spirit, it can affect our connection with animals because our pets rely on us for guidance and leadership. If we are out of balance, they will feel it. Identifying what fulfills you and taking action to feed your spirit will clear and strengthen your energetic channels for communication with animals (and everyone, for that matter).

My main advice for maintaining a healthy spirit is to make time for your outlets and passions. When your spirit is high, you have more energy, more motivation, and feel more fulfilled. Whether it's singing, kickboxing, spending time with close friends, dancing, gardening, or anything else that fills you with joy and peace, just do it. Schedule it into your calendar if you need to!

Variety is also an important key to feeding your spirit. For example, when I was working with Linda, I could feel through her dogs how she was not feeding her spirit. They showed an image of what felt like her in a fog: everything was gray and the walls were closing in on her from all directions. At first, she was very defensive when I shared this with her and she replied, "My rescue work feeds my spirit!" Maybe at one point it did, but at this point, it was sucking the life out of her. Attending to the dogs consumed her; she hadn't even showered in four days when I visited. So, I asked her what she loved to do before she opened the rescue. She shared that she loved to dance, but she had no time or energy for that anymore. She also told me she had a huge collection of movie memorabilia—boxes of rare posters, autographs, costumes, and other items that she had lovingly acquired through auctions and estate sales. But all of that was boxed up in her garage because her house was filled with dogs. There was no place to display everything so she could actually enjoy it.

I noticed that she had a spare room in her house that was cluttered with old dog crates and supplies, most of which were not usable. I suggested she donate or discard what she no longer needed and use the space to display all of the memorabilia items that she loved so much. I also encouraged her to make time to dance. During our follow-up session, she surprised me with the transformation of her spare room into what she called her "spirit room." She had painted the walls vibrant colors and displayed all of her memorabilia. There were a few cozy spots to sit where she could meditate and appreciate everything around her. She had also signed up for a dance class and scheduled time every other weekend to go out dancing with her friends.

On this visit, Linda's energy was completely different. There was a twinkle in her eyes, rosiness in her cheeks, and a wonderful spring in her step. She told me that she felt that her animal communication skills were improving with her rescue dogs and there were fewer challenges managing them. I could feel in all of the dogs that they saw her vibrating with a technicolor kind of energy, very much in contrast to the gray energy they were seeing before. The dogs shared with me a feeling of there being more room to breathe without the intensity and pressure that previously filled the house. Linda was helping herself feel more balanced by feeding her spirit. And the dogs benefited because she was clearer in her communication with them and more grounded in her guidance and leadership.

Make tending to your spirit an item on your checklist so it doesn't get forgotten among all the things you have to get done. Your well-being depends on it, and your animals depend on you to be balanced and fulfilled.

Intuition

Intuition is a form of instinctive knowledge or perception that arises without the need for conscious reasoning or analysis. It can be experienced as a gut feeling, hunch, or sense of inner knowing beyond rational explanation. Intuitive experiences can take many forms and may vary from person to person. Intuition is often referred to as the "sixth sense" because, just like sight, smell, hearing, taste, and touch, it helps us to perceive and interact with the world. It keeps us on track and guides us on our journey through life.

Have you ever been thinking about an old friend and then they suddenly call you out of the blue? This happens to me consistently with my friend Rachel, whom I have known since I was five. We don't talk often because our lives have taken different directions, but whenever she pops into my head, I soon get a call from her. I always answer with, "Oh my gosh—I was just thinking of you! Of course, you called!"

Other examples of an intuitive experience could be a sudden feeling of danger or unease that leads you to avoid a certain situation or location where there is an accident. Maybe you've had a strong sense of familiarity or recognition when meeting someone for the first time, or a feeling of connection or resonance with a particular place or object. Or maybe you've had a vivid dream that provided you with intriguing insight. An intuitive experience could also be a feeling of empathy and understanding for someone without knowing their full story. You may experience a gut sense about how an animal is feeling, what they need, or what they are trying to tell you.

Recognizing Guidance

Intuition is something we all have access to, and it becomes stronger and more reliable with regular exercise. When you tap into your intuition, you become more aware of your inner guidance and more aware of subtle cues from the world around you. You'll be more sensitive, empathic, and even make better decisions.

Our intuition is always tuning in to the messages around us. One time I was driving home from the grocery store, and I glanced at a billboard that said, "Reconsider your options." Seeing these words sparked a feeling within me to suddenly take a left turn, even though it was out of the way. As I drove through a random residential street, I saw a small white poodle in the middle of the road. I got out of my car and approached the friendly dog. I called the phone number listed on the name tag, and the guardian was shocked to hear that their dog had escaped their yard, six blocks away. I drove the dog back to its home, and the woman was eternally grateful. As I continued home, I saw a car accident that was right on the path of my usual route home, but my diversion helped me avoid it. I recognized the power of my intuition and expressed my gratitude.

Intuitive guidance may come to you anytime and anywhere. Therefore, it's important to record it when you receive it. If you don't have a pen and paper, use a notes app or voice memo on your phone. Don't just assume you will remember it. I often receive an inspired idea or important action step while doing something else, so I shoot a quick text message to myself. Later when I check my phone, I often have no recollection of recording that intuitive guidance and am so grateful I preserved it. We are thinking and feeling so much all the time, these subtle intuitive messages may slip by. Even though in the previous story the guidance came from a huge billboard, if I had been stewing about bad traffic or thinking about what I would make for dinner, I may have missed it or ignored that gut feeling to turn. You also never know when a million-dollar idea may subtly slip into your consciousness!

If intuition feels like it is difficult to hear or feel, it is just that you need to rebuild that connection. Set an intention to be open and curious to intuitive guidance and allow your imagination to support you. Become receptive to any signs, symbols, and messages, and be willing to be flexible in your interpretation and understanding. The rational mind may try to minimize or discredit your intuitive hits. It is easy to just think it's all in your head. But when you feel like you are receiving a message, say it out loud and own it. Then notice and appreciate when your intuition is correct. The more you acknowledge when your intuition is working, the more awareness and clarity you will have as more guidance comes to you.

The Four Clairs

Intuition can be experienced in multiple ways. The primary ways are often known as the four psychic senses, or four clairs:

- *Clairvoyance*—the ability to see things in your mind that are not visible to the physical eyes, such as visions, images, and colors. The *mind's eye* and *third eye* are often terms used when connecting to your intuition through the mind.
- *Clairaudience*—the ability to hear sounds, voices, or messages that are not audible to the physical ear.
- *Clairsentience*—the ability to feel or sense things beyond the physical body's experience, such as emotions, energy, or the presence of someone or something.
- *Claircognizance*—the ability to experience an inner knowing that is not based on physical evidence or logical reasoning, such as sudden insights or intuitive hunches.

Some people experience one of the clairs, and others experience a few or all of them. The more you become aware of how you receive guidance and acknowledge it, the stronger and more supportive your guidance will become.

Practice: Developing the Clairs

To explore the four clairs, you can engage in meditation exercises that focus on each of these intuitive abilities. For each, bring to mind a person, animal, situation, or place that you would like to tune in to energetically. Then, follow these suggestions for whichever clair you are working on:

- *Clairvoyance:* Visualize your subject in your mind's eye as vividly as possible. Use all of your senses to capture any images or visual impressions, such as colors, shapes, textures, sounds, and smells.
- *Clairaudience:* Focus on any sounds or words that you might hear in your mind's ear as you focus on your subject. Allow yourself to be open and receptive to any sounds that come through, even if they seem faint or unclear at first. Pay attention to the tone, pitch, and rhythm. Take note of any emotions or feelings that might be associated with what you hear.
- *Clairsentience:* Bring your attention to your body and notice any physical sensations you may have. Focus on any areas of your body where you are holding tension or discomfort and try to relax those areas. Then focus on your subject and notice any sensations or feelings that arise in your body. You might feel warmth, tingling, pressure, or other physical sensations. Pay attention to the emotions that arise in your body.
- *Claircognizance:* Bring to mind a question or situation that you would like to receive intuitive guidance on. It could be related to any area of your life. Allow yourself to become open and receptive to any sudden insights or ideas that come

through. Pay attention to any thoughts or ideas that pop into your head. Trust your gut instincts and allow yourself to follow any sudden hunches or ideas that come through.

Simply observe any impressions that arise, even if they don't make immediate sense. Try not to overthink or analyze what's coming through. To conclude, take a few deep breaths and allow yourself to come back to the present moment.

After, take note of any insights in your journal. You might ask yourself some questions to deepen your understanding of what you are sensing and feeling. For example, "What is the message that this energy is trying to convey to me?" or "What can I learn from this experience?" Overall, the key to developing your intuitive abilities is to practice regularly and trust your inner guidance.

The mind, heart, spirit, and intuition are four distinct yet interrelated internal components that help us understand and respond to our experiences. Having more awareness about each of these allows us to be in tune with the wider world and live more cohesively. The next chapter expands on this, helping us become better receivers of these energies and messages.

Journal Reflections

1. Picture your inner saboteur—your gremlin—clearly in your mind. Describe them and give them form. What are their motivations? What are they afraid of? What are they stopping you from trying? What is a better job you can assign them that will support who you are today?
2. Choose a day to keep track of the emotions you experience. From the moment you wake up, every time you feel some kind of emotion welling up in you, just jot it down without any analysis or

judgment. What happens when you recognize your emotions?

3. What feeds your spirit? How can you better honor your spirit, outlets, and creative expression? What action steps can you take?
4. Think of a time when you experienced a strong gut feeling. What was that experience like for you? Did you listen to it or ignore it? What was the outcome?

BECOMING A STRONGER RECEIVER

Now that you understand how to sense energy around you, you are probably getting familiar with how those energetic vibrations affect you. For example, the tension from an argument or the sorrow from watching bad news can lower your vibration. You absorb that energy, and it can stick with you, perhaps infiltrating other areas of your life and manifesting as a sick feeling in the stomach or tightness in the neck and back. The next question is, how do we control what happens to the energy we take in?

In this chapter, we will discuss how to clear blocks and energetic clutter, become an objective observer, and support spiritual health through physical health. These kinds of energetic self-care are important because they help us maintain a healthy and balanced energy system, which is essential for our physical, mental, emotional, and intuitive well-being. Because animals are sensitive to energy, they feel when we are weighed down by our thoughts, feelings, and emotions. So, with a balanced energy system, we are better able to tap into our intuition and more effectively understand and communicate with our animals.

Clear Blocks and Energetic Clutter

To be a stronger receiver of the energy around us and the information it offers, we must take into consideration what affects our energy, what might block or clog our energetic receptors, and what is adversely influencing our abilities to tune in to our environment, those around us, and especially ourselves.

What Is Draining You?

There are a few main reasons why our energy might be negatively affected. First, we may experience certain people, places, things, or animals draining our energy. Whether we are intentionally sharing our energy or not, some interactions just leave us feeling taxed. Have you ever spent time in a crowded mall or busy city street and felt suffocated? Or been stuck in bumper-to-bumper traffic for a long time and left feeling exhausted? Even walking through an animal shelter can be an emotional roller coaster due to the stress, anxiety, and compassion from all the animals, workers, and visitors filling the space. This isn't tied to space either; for example, observing negative energy through social media can also deplete us.

Next, we tend to function—and thrive—with routine. Routines create a supportive rhythm for us to follow and in some ways are unavoidable. Consider your average morning. I bet there are at least five things you do every day: walk the dog, make breakfast, brush your teeth, and so on. But because routines are the same repeating patterns, the energy may become stagnant. We may feel stuck or uninspired, which affects our vibration and the energy we share with others.

Finally, a cluttered environment can also drain our energy. Physical clutter, even if you are good at ignoring it, can be overwhelming to the mind. It is important to be mindful of keeping your living and workspaces clean, fresh, and organized. We may not realize why we are having challenges understanding our pets, when in fact, the cluttered home environment is creating an energetic fog that is affecting everyone.

So, recognizing what is draining your energy is the first step to restoring balance. Let's look at an example of this in action. Jenna worked as a vet technician in a very busy emergency animal hospital. During the height of the pandemic, her facility was struggling.

Employees were out sick. Several people quit. Pet adoptions were at an all-time high, which meant more animals needed health care. Safety measures such as curbside drop-offs meant more work for vet workers and more stress for guardians. Jenna had to help with responsibilities beyond her job description, such as answering phones and booking appointments, all while she was assisting the doctors during procedures, completing paperwork, handling animals, and communicating with guardians. One minute Jenna was dealing with someone yelling at her, the next minute she was consoling someone whose pet had passed away. She was working long hours six days a week.

She was committed to service and kept reminding herself why she was in this line of work, but it was getting the better of her. She would come home exhausted, often crying, and she was irritable with her husband. Her two dogs started to display separation anxiety and began chewing on their paws when she was home. Her home became cluttered: stuff needing to be organized or thrown out, laundry and dishes that needed to be cleaned. She spent every free moment constantly checking the news on her phone to get the latest updates on the pandemic, and she would fall asleep each night watching the tragic news stories as they unfolded.

Jenna reached out to me for help, primarily because she was worried about her dogs and knew she was not communicating clearly with them. When I met with her, I could feel her carrying so much energy that was not her own. She didn't even realize how many thoughts, feelings, and emotions she was taking in from everyone around her at work. But her dogs were feeling it all, and it left them confused, stressed, and anxious. Things clearly could not continue this way.

This was an extreme case where Jenna's energy was being drained in almost every aspect of her life. She was drained by bad interactions with clients and her co-workers' stress. She was drained by worry for her dogs and their new behavioral issues. She was drained by the mounting clutter and to-dos in her home. She was drained by the globalized trauma of the pandemic.

I first recommended that she sit with a journal for a few minutes each day and check in with her thoughts and feelings, asking of each one, *is this mine*? Simply pausing to ask that question can shift our perspective so we can recognize when we are carrying energy that is not our own. Jenna journaled about the animals she was helping and what they experienced during their hospital visits. She wrote about the people who brought their pets in—how they cried, yelled, sat in tense silence. She wrote about the stress she observed in the vets she assisted during operations and the pressure on them to hurry up because there was a long line of other emergency cases. With each person and animal that came to mind, she gave a voice to what she felt were their thoughts, feelings, and emotions.

She also journaled about how this made her feel, the frustration and sadness that she was exhausted and tense at home with her husband and dogs. She wrote about how her energy seemed to be affecting them. She wrote about how suffocated she felt by all the piles of laundry, dishes, and clutter that were amassing because she did not have time to deal with it all. Just writing about it made it hard for her to breathe deeply. But as she read what she had written, she realized that all of it was not related to her. She had been absorbing other people's energy and assuming it was her own. This exercise was a huge breakthrough moment for her and gave her the clarity to recognize the energetic blocks and clutter that needed to be cleared.

Energetic Cords

Energetic cords are the invisible energetic connections or ties that exist between people, animals, places, and things. These cords are made of subtle energy and can carry emotions, thoughts, and intentions. Energetic cords can be both positive and negative, depending on the nature of the relationship, and affect our energy levels, emotions, and overall well-being. Positive cords can foster feelings of love, connection, and support, providing us with support and nourishment. Negative cords can create feelings of attachment, dependence, and toxicity, leaving us energetically depleted.

Energetic cords can form between individuals who have close relationships, such as family members, pets, romantic partners, friends, and even co-workers. They can also form between individuals and places, such as our homes, workplaces, or favorite vacation spots. Additionally, energetic cords can form between an individual and their own thoughts, beliefs, and emotions.

Now that Jenna understood what blocks she had, I taught her the following process of cutting her energetic cords. Jenna incorporated this into her new morning meditation practice. She began to feel lighter and more herself. She had more energy and patience, and the awareness to recognize when she needed a break. While sitting in her meditations, she started to receive new ideas about how to organize socially distant patient check-ins that were sensitive to the needs of their customers. She also received an intuitive nudge to do some searching online for new client management systems that would help process invoices and paperwork. She felt so inspired by these new ideas that they uplifted her energy.

Practice: Cord Cutting Meditation

Cutting or releasing energetic cords that no longer serve our highest good can be super helpful in restoring balance as a method of self-care. This can be done through various techniques, such as energy healing, but this visualization practice is a great place to start. By releasing negative cords and strengthening positive ones, we can cultivate healthier and more fulfilling relationships with ourselves and others.

- Find a quiet and comfortable place to sit or lie down, where you can relax and focus without distraction for about 10 minutes. With your eyes closed, take a few deep breaths, inhaling through your nose and exhaling through your mouth.
- Visualize your body in your mind's eye and scan around it in a clockwise motion. As you scan, feel the different parts of your body and their energy to get a sense of any energetic cords attached to you.

- Visualize any cords of energy that may be connecting you to the person, situation, or energy you want to release. See the cord clearly in your mind's eye, noticing its color, shape, and texture.
- Take a moment to acknowledge any emotions that come up for you as you focus on the cord. Allow yourself to feel whatever you need to feel, without judgment or resistance.
- When you are ready, imagine that you have a pair of scissors or a sword in your hand. Visualize yourself circulating your body, cutting the cords of energy with one swift and decisive motion.
- As you cut the cord, repeat a positive affirmation to yourself, such as "I release all negative energies and attachments from my life," or "I am free and empowered to create my own path." Circulate your body as many times as you feel guided to cut all the cords.
- After you feel complete, circulate your body again and fill each corded spot with light, as if you are gently patching holes in a wall.
- When you're ready, gently open your eyes and awareness and come back to your space.

Energy Shielding with Light

While cutting energetic cords helped Jenna release that built-up negative energy, she still had to go to her stressful and high-intensity job every day, which meant that energy would still be swirling around her. So, for the end of her daily meditation, I taught her the following energy-shielding techniques.

Energy shielding is an accessible way to protect yourself from intrusive energies so you can maintain a higher vibration. While there are lots of ways to go about this, I have found shielding with light to be very powerful for myself and my clients. Working with frequencies of light can strengthen our connections to higher consciousness, spiritual awakening, and transformation. When experienced, it can feel like a sense of warmth, radiance, or expansion in the body and energy field.

Practice: White Light of Protection

White light is the strongest energy and light of protection. It can block anything from blending with your frequency and adversely affecting you. White is what we see when all the wavelengths of light reflect off a surface; it directs its energy outward, so surrounding yourself with this color scatters energy away from you. The white light of protection will help if you are entering an intense situation or when you feel threatened and want to block out anything that could challenge your safety, security, and balance.

1. Find a quiet and comfortable place to sit or lie down, where you can relax and focus without distraction for about 10 minutes. With your eyes closed, take a few deep breaths.
2. Visualize a bright, glowing white light surrounding you. See the light as a powerful shield of protection, surrounding you in a bubble of positive energy.
3. As you inhale, imagine that you are drawing this white light into your body, filling you with its positive energy and shielding you from negativity.
4. As you exhale, imagine that any negative energy or emotions within you are leaving your body and dissolving into the light.
5. Continue to inhale the white light and exhale any negativity for several minutes, feeling the light surrounding and protecting you.
6. When you feel ready, visualize the white light expanding to surround your loved ones, animals, home, and community, spreading positivity and protection.
7. When you're ready, gently open your eyes and awareness and come back to your space.

Practice: Pink Light of Support

The color pink is associated with the heart chakra, an energy center in the body that governs our ability to give and receive love. Pink light will support you in keeping lower vibrations from affecting you, while still allowing good, higher vibrations to be received. For example, you may be interacting and engaging with an animal who you want to communicate with and have a sense of their feelings, but you do not want to take on their energy. The pink light allows you to connect with them and empathize without being energetically affected by any lower energetic vibrations.

1. Find a quiet and comfortable place to stand, sit, or lie down, where you can relax and focus without distraction for about 10 minutes. With your eyes closed, take a few deep breaths.
2. Visualize a ball of pink light in the center of your chest. See the light growing brighter and more vibrant with each breath.
3. Imagine the pink light expanding outward, filling your entire body with a warm, gentle glow. See the light radiating from your heart center and spreading out to your arms, legs, head, and feet.
4. Next, visualize the pink light expanding beyond your physical body, forming a bubble of loving and healing energy around you. See the bubble growing larger, extending several feet in all directions, encompassing your entire aura.
5. Repeat a positive affirmation to yourself, such as "I am surrounded by the healing and loving energy of the Universe," or "I am open to giving and receiving love and compassion."
6. Take a few deep breaths and allow yourself to bask in the warm, comforting energy of the pink light. Imagine it enveloping you in a cocoon of love and support.
7. When you're ready, gently open your eyes and awareness and come back to your space.

With all the spiritual self-care Jenna was doing, she began noticing she had a calming effect on the people and animals at the hospital, and the vets that she assisted kept thanking her for being such a peaceful and grounding energy in times of chaos and stress. They recognized how relaxed animals became when Jenna was in their presence. She also saw a huge shift in her dogs at home. They were less needy and no longer chewed on their paws with anxiety. Her relationship with her husband improved too. They began to make time for date nights at home with delivered food and a movie, and they would go for bike rides and family walks together with the dogs. Jenna was able to successfully clear her blocks and understand what was her energy and what was not. She was more in tune with what energy she could bring to the moment that was serving the highest good for everyone—herself, her husband, her dogs, her co-workers, her customers and their animals, and the world.

Become the Objective Observer

When tuning in to our environment and others around us, it's important to assume the perspective of the objective observer. This means observing the world around us, as well as ourselves, without judgment or attachment. It is the practice of being fully present in the moment and witnessing our thoughts, feelings, and experiences without getting caught up in them. As the objective observer, we can step back from our biases and see things as they truly are. We detach from our emotions and those of others and can view situations neutrally. This practice can help us gain insight into ourselves and others, including animals. It can also help us cultivate compassion and empathy. By developing this practice through meditation, mindfulness, and self-reflection, we learn to detach from our egos and see the world with greater clarity.

Our ego is the part of us that is concerned with our identity and sense of self. It is the part of us that wants to be recognized, validated, and admired. The ego is characterized by qualities such as self-importance, self-centeredness, and a need for control. Perceiving objectively helps us to align with our true self, the part of us that is connected to a higher power and our intuition.

Often, when we observe an animal, another person, or a situation, our feelings and opinions influence our interpretation. For example, I may be visiting someone and their cat, who has terminal illness and looks visibly unwell. The human may also be emotionally overwhelmed by the situation. I might feel horrible seeing the cat in such a suffering state, and I may feel sad for the human as they cope with this. But to get a clear line of connection and communication with the cat, I must put aside my feelings. So, to accurately assess through our intuition, we detach from our subjective thoughts, feelings, and emotions so our perceptions are not influenced and skewed. Setting an intention of being the objective observer and letting go of any personal emotions that arise will help you maintain a balanced exchange of information.

Let's refer back to Jenna. I shared with her the process of being the objective observer so that she could put it to work at the vet's office. Before, she had no energetic boundaries and was always empathizing with the animals at the hospital. Quite often, animals needed treatment due to human negligence. She shared a story about a family's labradoodle that had a habit of ingesting items the family left lying around, including toy dolls and Christmas ornaments with metal hooks. The dog had to go through major surgery to remove the ornaments, and it was the third time the dog had been brought in for this issue. The previous two times, Jenna felt extreme anger toward the guardians. She felt so sorry for the dog and had to step outside to cry after seeing the dog struggling and suffering. It took everything within her to hold herself back from screaming at the people, "What the heck is wrong with you?! How are you letting this happen?"

But after she learned how to be the objective observer, she was able to detach from the situation. She was able to do her job with ease and success, supporting the doctors in the removal surgery. She felt confident in her ability to intuitively communicate with the dog before, during, and after, sharing calming and soothing energy to help them feel safe. Jenna was also able to objectively observe the people without judging them and offer suggestions to make sure it never happened again. She felt that her messages and guidance got through to them more clearly since she was able to project centered calmness, and there has not been another incident since. The family was very grateful to Jenna for her kindness and suggestions on how to better care for their dog, who they loved so much.

Practice: Objective Observer Meditation

With this practice, you will learn to become the objective observer of your thoughts, emotions, and experiences. As you detach from your own biases and judgments, you will gain greater insight and understanding into your own life and the world around you.

1. Find a quiet and comfortable place to sit or lie down, where you can relax and focus without distraction for about 10 minutes. With your eyes closed, take a few deep breaths.
2. Without judgment, begin to observe your thoughts, feelings, and emotions. Let them come and go, without getting attached to them. You may notice them and say to yourself, ***That's interesting***. Continue to let them come and go with this perspective.
3. If anything is feeling intense, distracting, or heavy, ask yourself, ***Is this mine?*** You may feel some are not your own. Release them. If you'd like, you may place them in a bubble and see them float away into the atmosphere.
4. Observe the sensation of your breath as it enters and leaves your body. Notice any physical sensations, such as discomfort or relaxation. Observe them without trying to change them. Again, ask yourself, ***Is this mine?***
5. If your mind wanders, simply notice the thought that pulled you away and then bring your attention back to your breath.
6. As you practice, you may begin to notice patterns in your thoughts, emotions, and physical sensations. Observe these patterns without getting attached to them.
7. When you're ready, gently open your eyes and awareness and come back to your space.

Support Spiritual Health with Physical Health

Through our bodies, we are connected to the world around us, experiencing life, relationships, and the natural world. Just as a temple is a sacred space set apart for worship and connection with the divine, the body is a sacred vessel that is meant to be respected and cared for.

Physical health plays a significant role in supporting spiritual and intuitive health. When we take care of our physical bodies, we experience many benefits that positively impact our spiritual and intuitive well-being. Regular exercise, healthy eating habits, and adequate sleep give us more energy and help us feel more in tune with our intuition.

Physical health can also help us build emotional resilience, or the ability to bounce back from challenges and setbacks. When we feel good, we are better equipped to manage stress, anxiety, and other emotional challenges. Our physical health can also support our connection to animals and nature. When we engage in physical activities like hiking, gardening, walks along the water, or simply spending time outdoors, we feel a greater sense of connection to the natural world and our spiritual selves. This will calm and clear our energy body and elevate our energetic vibration. Animals feel and respond positively to these vibrations because they instinctually are drawn to what feels good.

Exercise

When we are not physically moving very much, we are susceptible to energetic blocks. On the other hand, when we move our body, our energy flows: we take in new energy with each inhale and release old energy with each exhale. Over the years, I have learned that I need to take care of my physical body to feel mentally and emotionally balanced and connected to my intuition. It was the same for Jenna, who, in pre-pandemic times, had a daily running practice. But when work became overwhelming, she felt she didn't have time to exercise anymore. Self-care is often the first thing to go in times of stress, but that is when we need it most! When Jenna made it a priority to get outside and run again, she quickly felt more mentally clear. That clarity helped her meditate, which led to huge strides at work and in her personal life.

I regularly get outside to breathe in fresh air and get my blood pumping. I especially love hiking with my dogs, taking in the beautiful views and just being among the trees, rock formations, and the energy of nature. Movement helps me release that old energy so I can keep my vibration high for my work with animals and their people. Cardiovascular exercise is especially helpful in creating energetic balance because it involves continuous, rhythmic movement that increases heart rate, respiration, and circulation. It doesn't matter what you do, so find something you love—skateboarding, jump roping, jogging with your dog—and make it a habit.

I also stretch my body out every day. Stretching improves flexibility not only physically but also energetically and mentally. When we are stagnant and stationary, our muscles constrict and tighten, which is painful and causes energy to become stuck in our body. This lack of flow and circulation leaves us feeling blocked and negatively affects our awareness. But when we stretch, we loosen our muscles and remove those energetic blocks. Yoga is a great way to get your stretch on, but you can also incorporate simple stretching throughout your day. Often when I've been sitting for long periods, I feel drained of energy and my brain is scattered. But by simply standing up, stretching my body out, and taking some deep breaths, I instantly feel energized and restored.

Nutrition

You don't need me to tell you that a balanced diet rich in nutrients nourishes the body and supports overall physical health. It also boosts energy levels and mental clarity, which can make it easier to connect with your intuition.

Don't get me wrong—I have a huge sweet tooth and love cookies, cakes, and pies. I am all about enjoying life and doing what feels good! But I make it a practice to notice how different foods and drinks affect my energy. There is life-force energy in everything we consume; some foods elevate your life-force energy and some slow you down. Processed foods and sugary drinks dull my energetic frequency and cloud my mind, whereas fresh fruits and vegetables, especially anything green, amplify my intuition. For me, moderation is key because too much of a good thing throws my body and energy out of whack. And when I indulge, I have learned that a fresh, homemade cookie does not adversely affect my energy like packaged cookies do.

I am always drinking water too, as it's an essential component of the human body, playing a vital role in regulating body temperature, transporting nutrients and oxygen to cells, and keeping us hydrated. As soon as I wake up every morning, I drink a 16-ounce glass of water with freshly squeezed lemon. Cucumber, lime, mint, and basil are also wonderful flavor additions. I also recommend carrying a 32-ounce water bottle with you to help you remember to drink water throughout the day. You may even consider using your phone or a timer to remind you to drink water at regular intervals. This can help you establish a routine and make drinking water a habit.

There was a brief time when I was not making time for exercise and proper nutrition due to my schedule. I was getting a lot of inquiries and felt that I needed to help as many people struggling with their pets as I could. I was overscheduling myself and no longer making time for workouts or hikes with my dogs. I was also cutting corners in my diet by skipping meals and eating protein bars between sessions. This took a toll on me very quickly. I was feeling drained and irritable after every full day of client sessions, and I was more susceptible to illness too. I was tired and moody, and my body felt tight and tense. To feel balanced and capable of serving people and their animals, I realized I needed to make time for my health.

As soon as I reorganized my calendar and blocked out time each day for exercise and proper eating, my energetic vibration elevated and my health improved. I felt stronger in my sessions and more intuitive information flowed through me, so I was able to better help my clients. Exercise, stretching, and healthy eating keep my battery charged and my intuition on point; these three are nonnegotiable necessities for me.

Being a strong intuitive receiver is really all about self-care. It is more than being able to notice energy; it is the ability to shield yourself from negative energy and cut cords if necessary. It is detaching yourself emotionally so that you can assess situations objectively. And it is building up your inner strength and resilience through physical well-being. Now that you have an understanding of why it's important to cultivate calm, clear, and confident energy as well as how to go about it, you are ready for Part II, where we will jump into animal communication!

Journal Reflections

1. Sit quietly for a minute and tune in to what is occupying your vibrational space. What feels good? What feels distracting? Feel into the essence and energy of everything around you, tuning in to specific things and describing what you feel using all your senses. After, take some time to journal on what you felt—what energy belonged to you? What didn't? What can you release?
2. What does being an objective observer mean to you? Try being the objective observer in any situation in your life, detached and releasing judgment. Describe your thoughts and feelings about the experience.
3. Choose a day, week, or even month to pay attention to how the foods and drinks you consume make you feel. After you eat or drink something, make notes in your journal about how you are affected. Do you notice any changes in your thoughts and emotions? How does your body feel? Is your intuition guiding

you toward certain foods and away from others? Ask your inner guidance to help you understand what foods and drinks support you and what is not serving you.

4. Tune in to your physical body and ask yourself, *How flexible am I, physically and emotionally? How is my energy flowing?* What changes can you make to be even a little bit more flexible? Write these down and commit to them.

Part II

PROCESSES, TOOLS, AND TECHNIQUES

UNDERSTANDING ANIMAL COMMUNICATION

Now that we have created a foundation of and practice for calm, balanced, and centered personal energy, we are ready to start incorporating intuitive animal communication. Broadly, *intuitive animal communication* is the practice of using telepathic, energetic, and intuitive abilities to communicate with animals on a deeper level. This practice is based on the premise that all living beings, including animals, have consciousness and can communicate on a nonverbal, energetic level through thoughts, emotions, images, and sensations.

As we have been learning, an intuitive connection goes beyond observable behaviors and body language. For animal communication, it involves tuning in to the animal's energy and emotions, so we are better able to understand and care for our animal companions. I often describe it as feeling a sense of empathy or emotional resonance. We can use intuitive communication to gain insight into an animal's personality and purpose, resolve behavioral issues, address health concerns, and establish a deeper connection and bond. We may be able to pick up on the animal's emotional state, even if it's not obvious from their behavior or body language. Sometimes this is through what they share with us, and other times we may sense something going on within them, even if they are trying to tell us otherwise.

In this chapter, we will focus on background information, such as about different kinds of communication, communicating with different types of animals, and utilizing the tools you have been developing throughout Part I specifically for work with animals.

Animals and Telepathy

To create a deeper connection and understanding with animals, we first align our energy and vibration with them, which is known as becoming a vibrational match (see Vibrational Match Connection, page 119). This involves tuning in to the animal's energy and emotions and raising our vibration to match theirs. When we become an animal's vibrational match, we begin to truly communicate with them on a deeper level through telepathy or other nonverbal means.

The practices in Part I about meditation, mindfulness, and energy work are the foundation for aligning vibrations. By cultivating a deeper awareness of our energy and vibration, we can become more attuned to the energy of the animals around us and connect with them more profoundly. Becoming a vibrational match with animals can also have healing benefits mentally, emotionally, spiritually, and sometimes even physically, both for the animal and ourselves. By aligning our energy with that of the animal, we can help them release any negative energy or emotions and promote a state of balance.

A note about telepathy, as that term may seem complex or advanced: put simply, telepathy is the ability to communicate information from one mind to another without the use of physical senses or conventional communication methods. The word *telepathy* comes from the Greek words *tele* meaning "distant," and *pathos* meaning "feeling" or "perception." Telepathy is often considered a form of extrasensory perception (ESP) or a psychic ability and is believed to be related to the energy and vibrational frequencies of the human mind and consciousness. It is commonly described as a mind-to-mind connection in which thoughts, emotions, or images can be transmitted between individuals without the use of spoken or written language.

I believe—and have seen with my clients—that everyone can tap into telepathic skills. But, just like any other skill, it is something that requires consistent and dedicated practice. Some people experience intuitive communication with animals here and there, sometimes feeling they are receiving a message or a "download" unexpectedly and out of nowhere. But it may also be experienced randomly, without the person having clear control of it or confidence in what they are experiencing. As a child, this was how I first experienced intuitive animal communication.

When I was around 10 years old, an orange tabby cat I had never seen before crossed my path down the street from my house. She rubbed her body against my legs, and when I leaned down to pet her, I saw an image in my head of a small red house, almost like an old schoolhouse. The red paint really struck me for some reason. About a week later, I was over at the house of a new classmate who just moved in down the street from me, and there was the orange cat. While we played in the backyard, I saw a red garage that was shiny from a fresh coat of paint and it had a big bell on the outside, like an old-fashioned school bell.

I told my new friend that I saw this red garage in my head when I met his cat on the street a week prior. He said that didn't surprise him because the cat had been scared of the painters and disappeared for a day. He also said that when they found the cat and brought her home, she wouldn't go anywhere near the garage, and they thought it was because of the smell of the new paint.

When I encountered the cat, I had not made a proactive effort to communicate with her; she just communicated with me. It was that simple. In this way, some animals may be easier to communicate with than others. Especially if you are giving off a calm and high energetic vibration, you may find that some animals even seek you out, like that sweet orange tabby.

Connect with Love, Joy, and Gratitude

Love, joy, and gratitude are powerful emotions that can help amplify our connection and communication with animals in several ways. When we experience positive emotions, we radiate positive energy and openness. This energy can be sensed and felt by animals, who are often highly attuned to our emotional states. When we approach animals with positive energy, they are more likely to feel comfortable and at ease around us, whether it's remotely or in their physical presence. These emotions can also foster trust between humans and animals and deepen our empathy as we are more likely to see them as sentient beings with their own desires and feelings.

To cultivate positive emotions when interacting with animals, start by focusing on the positive aspects. This could involve expressing gratitude for the animal's presence, finding joy in their good qualities, or feeling a deep sense of love and compassion for their unique spirit. Cultivating the energy of love, joy, and gratitude and directing it toward energetically supporting animals can be a powerful way to enhance communication.

Here are some ways to go about this:

1. *Send positive intentions,* creating a powerful energetic field that supports an animal's well-being. Visualize a field of loving, joyful, and grateful energy surrounding the animal, and intend for this energy to nourish and uplift them.
2. *Practice energy healing* through techniques such as directing a high vibration life-force energy (see Cultivating Life-Force Energy of Nature Meditation on page 87) can be used to direct positive energy toward animals, balance their energy field, and support their physical and emotional health.
3. *Engage in prayer or meditation* to create powerful intentions that can have a positive impact on animals' energetic fields.

4. *Use crystals and healing stones,* such as rose quartz or amethyst, which are believed to have a calming and soothing effect on animals. If we feel a connection with and power in specific crystals and stones, the animals will feel these positive energetic vibrations. These can be placed safely near the animal's sleeping area or carried with you when interacting with animals. Note: make sure the crystals are out of reach of your animal, especially if they tend to chew on or eat on household objects.
5. *Spend time in nature,* which helps us connect with the natural rhythms and life-force energies of the earth and all living beings.

To illustrate some of these tactics and see how even small shifts in our energetic state can have a profound impact on animals, let's meet Catherine and Kaya.

Catherine contacted me because she and her family adopted a dog named Kaya from a meat trade dog rescue in Vietnam. In various countries, such as China, South Korea, Vietnam, and others, dogs are often raised on farms or captured from the streets and then sold for their meat, which is consumed by humans. Many organizations save dogs from the meat trade and Catherine's family chose Kaya through a listing on a rescue's website. Kaya was flown to California with a broken leg and a serious skin disease. Catherine's family paid for her leg surgery and skin treatment, and they welcomed her into their home two weeks later.

Needless to say, Kaya was extremely nervous after living in horrible conditions for the first two years of her life. She had just traveled across the world and was recovering from her medical issues—of course she was anxious. She was constantly trembling and was very fearful of Catherine, her husband, and their two young daughters. Catherine asked if I could come to their house to help. She wanted to know what Kaya needed and to communicate to her that she was safe and in her forever home. But along with Kaya's trauma, the whole family was experiencing their own forms of PTSD just from hearing Kaya's story and now living with her. The kids would regularly break down crying because they felt sorry for Kaya. Catherine would wake up in the middle of the night from nightmares that played out Kaya's story over and over. Catherine's husband was suffering from intense chronic headaches due to the stress of the whole situation.

I suggested to Catherine that we meet outside in their beautiful backyard. It was a bright, sunny day, and Kaya was not walking very well yet, so they helped her lie down on a comfortable, colorful tapestry. I knew Catherine and her daughters loved crystals, so I brought some clear quartz and amethyst and placed them around the edges of the tapestry. Surrounding ourselves with beauty can have very profound healing effects because it lifts our spirit and energetic vibrations. In this case, anything that would help uplift everybody's energy was necessary because there was so much sadness and worry in their home. Catherine and her daughters lit up when they saw the crystals, which I knew would support and surround Kaya with high energetic vibrations. We were already off to a great start.

We all sat in a circle around Kaya, and I asked them to share with me all their feelings of love, joy, and gratitude for Kaya as a member of their family. The tall, lush trees and blooming rose bushes that surrounded us made it feel as if we were supported by nature. As I listened to the family share their love for Kaya, I set an intention of absorbing these energies through every cell of my body. I also imagined I was absorbing all the life-force energy from the natural elements surrounding us, taking it all in and blending it together.

I then gently began to pet Kaya with the palm of my hand for several minutes as she lay on her side, calmly smoothing her fur and imagining I was sending her all this love and healing energy through my hand and into her body. In my mind's eye, I shared with her simple communication that all was well, she was safe, and she would forever feel this love and care. I felt back from her the sense of a flower slowly sprouting and beginning to open up, and that this process had begun the moment she arrived at their home. I also felt from her a feeling of her strength, desire, and motivation to do her part. It was as if she was saying, "I have it in me, believe in me, there is something more within me, and I want to be that dog."

I shared these messages with Catherine and her family, and it made them so happy to know Kaya felt accepted and at home. I suggested to the family that they practice this process regularly to continue to support Kaya's return to wellness. The next day, Catherine contacted me to let me know that Kaya was standing and walking a bit! And she was allowing everyone in the family to approach her. Her skittishness was gone and, sure enough, she was doing her part to move forward through her healing and adjustment to her new life. Sometimes simply sharing energies of love, joy, and gratitude can be powerful to open up the lines of communication and connection.

Practice: Cultivating Life-Force Energy of Nature Meditation

This meditation helps us tune in to the powerful life-force energy of nature that is very healing, supportive, and transformative. And it is everywhere and always available to us—in the ground, in the air, and within all living things in nature. We can call upon this energy to absorb it, cultivate it, and direct it for the highest good for ourselves and others. This process will help you experience a deeper sense of harmony with the natural world.

1. Find a quiet and peaceful spot in nature where you feel safe and comfortable and can relax without distraction for about 10 minutes. This could be in a backyard or nearby park, beside a stream, beneath a tree, or even just on a small patch of grass.
2. With your eyes closed, take slow, deep breaths—in through your nose and out through your mouth. Maintain this breathing throughout.
3. Turn your attention inward. Visualize a beautiful, natural scene in your mind's eye. This could be a forest, beach, mountain, or any other natural setting that you feel drawn

to. Imagine yourself standing in the center of this natural scene. Feel the energy of the natural world and allow yourself to become fully present in the moment.

4. Imagine that you are breathing in the life-force energy of nature. Visualize this energy flowing into your body and every one of your cells with each inhale, filling you with vitality and strength. As you exhale, imagine that you are releasing any negative or stagnant energy from your body. Allow yourself to let go of any tension or stress you feel.
5. Focus on your connection with the natural world around you. Imagine that you are a part of this ecosystem and that your energy is in harmony with the energy of the plants, animals, and other beings around you.
6. Spend a few minutes in this state of meditation, breathing deeply and allowing yourself to connect with the life-force energy of nature.
7. Imagine you are harnessing the life-force energy and sending it to the animal of your choice for healing and support. Visualize the animal gently surrounded by this energy, receiving and absorbing it within their entire body and being. When you feel they have received what they need, slowly cease the energy transmission.
8. When you are ready, gently open your eyes and take a few moments to ground yourself in the present moment. Allow yourself to carry this high vibration life-force energy of nature and share it with others as you go about your day.

Remote vs. In-Person Animal Communication

When you are first starting, you will likely be working with the animals you live with or others around you. But something that is sometimes surprising to people I work with is that remote (or distance) communication is equally effective and accessible. I should also note that when I intuitively communicate with animals (whether remotely or in person), it is all nonverbal. I am not talking aloud to the animal. Occasionally, I may share a word or two, but it's pretty much all through telepathy, energy, and intuition. Let's take a look at these two types of communication and some of the pros and cons.

Communicating Remotely

Communicating with animals without being physically present with them is based on the understanding that everything in the Universe is interconnected and that thoughts, emotions, and energy can be transmitted across distances. If you have been practicing the Energy Exploration meditation from page 33, you will certainly know this to be true!

No matter how close or far away from the animal we are, we can tune in to their energy and pick up on information about their emotions and overall well-being. This process can be done using a photo or video. This may be different for different people, but for me, a photo of the animal allows me to find and connect with their energetic frequency. For many years before video calls became a norm (and even to this day), I conducted sessions over the phone. The client would simply e-mail me a photo of their animal before the session. I also do many sessions through live video, and the animal does not have to be on the video screen with the person. Though I do not use this method, holding an object that belongs to the animal is another way to establish a connection.

It is also possible to not use any photo or video and just set an intention to connect with the animal with some basic information, such as a name, age, species, and breed. But I have found this method takes a lot longer; it can almost be like searching for a needle in a haystack. I find that a photo or video along with their name and age amplifies the process and helps me receive more accurate and insightful information. Being able to see the image of the animal gives me the opportunity to visually observe the radiance of their spirit and this swiftly aligns me with their energetic frequency for communication.

When I first started out doing sessions with people and their pets, I found phone sessions to be very productive because I could close my eyes and not have to worry about the person staring at me. There were fewer distractions, and I felt that I could tune in more deeply. But with time and experience, it became much easier and quicker for me to tune in, so now if someone is staring at me through video waiting for me to relay their animal's information, it does not affect me. These days, most of my sessions are done through video, and I work with people and their animals all over the world. The animal does not have to be on camera during the session, provided I have a photo of them. Using the photo along with feeling the animal's energy through their human during the live video session is all I need to see and sense them in my mind's eye.

Communicating in Person

In-person animal communication allows us to be physically present with the animal, which may provide additional sensory information such as the animal's scent and physical movements. Animals communicate through body language, and our physical energy and presence when we are in their personal space may also affect them. Some people naturally have a strong energetic presence, which may be overwhelming for a sensitive animal. In many animal species, such as dogs, sometimes cats, gorillas, bears, and chimpanzees, direct eye contact may be interpreted as a sign of dominance or aggression, so being mindful of this is respectful. By avoiding prolonged direct eye contact with an animal, we show them respect and acknowledge their autonomy. By respecting the animal's communication style and working with their needs and boundaries, we can create a positive and safe interaction.

In-person communication may also be affected by distracting environmental factors such as noise or other animals. Sometimes there are energies in a person's home that feel heavy or stifling. Or maybe the temperature feels too warm or cold. There can be so much energetic stimulation that it may feel like sensory overload. This may cloud and distract us from our abilities.

I remember one visit at a client's home where there were several animals, multiple young kids running around, a television on upstairs, and food cooking in the oven. On top of that, the home was very stuffy and had stale energy. There was also a police helicopter hovering above the whole time due to a local traffic accident, and the phone kept ringing. Talk about sensory overload! But through consistent practice of what we discussed in Part I, such as meditation, grounding, cutting energetic cords, energy shielding, and being the objective observer, it is possible to tune out all distractions and still successfully communicate.

Communicating with Different Species

In my experience communicating with various species of animals, the methods of connecting are all the same. In other words, I do not have one method for dogs, and another for cats, horses, or ferrets. But different animals do have different energies and personalities, which may affect how they communicate energetically and telepathically. For example, dogs often have a more direct and outgoing energy, while cats may have a more independent and subtle energy. However, I have communicated with cats that feel very doglike and vice versa.

Different animals may also communicate using different frequencies or symbols (we will dive deeper into symbols in Chapter 6) depending on their species and individual personality. A horse may communicate through powerful images or sensations due to their experience of their larger physical size and energetic presence. Whereas a bird may communicate through intricate and subtle patterns of energy because of their energetic experience of flying and pecking. Cats experience life in a more mobile way than dogs as they can climb trees, jump on counters, and get on their humans' shoulders. Dogs, on the other hand, are often experiencing a connection with humans through a leash and frequent walks together. The way each animal experiences life and engages with the world will influence the thoughts, feelings, and emotions they share.

As you begin to practice communicating with different species, I encourage you to take note of how you experience them and any similarities and or differences you notice. For example, you may have presumptions about how you will experience a rabbit's energy. But then after communicating with them, you may recognize how certain qualities in their energy are similar to, say, a cat.

Eternal Self vs. Earthly Self

When I communicate with animals, I am usually communicating with their *earthly self*. The earthly self is the physical body, personality, and temporary existence in the material realm. This includes the animal's physical attributes, behaviors, and experiences in their lifetime. The earthly self shares their likes, dislikes, and day-to-day experiences. For example, I may be communicating with a dog and their earthly self shares what toys and foods they like or what activities are fun and which make them feel anxious.

Sometimes in a communication session, the *eternal self* of an animal will come through with messages that are more profound, often related to bigger life lessons. The eternal self of an animal is their spiritual essence, soul or spirit, and ongoing existence beyond the physical realm. This includes their connection to the divine, innate wisdom and intelligence, and spiritual purpose or journey. I have felt and learned that animals have a spiritual essence equal in importance and value to that of humans, and they have a spiritual purpose and journey of their own. This spiritual essence is eternal, existing beyond the physical realm and transcending the limitations of the animal's physical body and personality.

This concept of the earthly and eternal selves of an animal reflects the belief that all beings have both a physical and spiritual dimension, and that the spiritual dimension is an essential part of the being's nature and identity. By honoring both the earthly and eternal self of an animal, we can cultivate a greater sense of compassion and reverence for all living beings. Both aspects have energy and information to share.

Let's take a look at an example of how information from the earthly self and eternal self might come through. Monica booked a session with me because she was concerned about her Weimaraner, Max. Max was diagnosed with a rare form of cancer at five years old, and the doctors were unsure about his prognosis. Monica was moving forward with chemotherapy but questioned if it was too much for Max to go through. Monica loved to take Max on camping trips, and they road-tripped all over the country together. She wanted me to check in on him to see how he was doing after his last chemo treatment. A part of her felt it might be good for them both to go out on a road trip again to feed their spirits since it had been a long time. But she didn't want to push him if he was too tired.

When I connected in with Max and Monica over video, his earthly self shared with me how tired and uncomfortable he had been, and in my mind's eye, I saw red balls of fire in his body that appeared extinguished but still smoking, as if they were cooling off. My sense was these spots were the cancer cells that the chemo was treating, and that he was in the process of healing. Max also shared that he was tired but also sick of being tired—he wanted to get back to life and doing things. He shared with me feelings of joy and exhilaration about road trips with Monica. I told her I was seeing pools of water and dolphins, and I asked her if they had ever traveled to the coast together. She said yes, and the last time they took a trip together it was to Sea World and Max got to watch dolphins up close.

I also felt Max's eternal self coming through and communicating that there were larger lessons for them both to experience related to optimism, faith, and taking a chance when the odds were against them. Max's eternal self shared with me a bigger purpose for his health challenges. I saw blank pages turning in a big book, and as the page would turn, words would appear on the page. I asked Monica if this made sense, and she said that because Max's cancer was so rare, the doctors were trying some experimental treatments. They were unsure about how everything would pan out, but Max was doing so well that they were recording their findings as a case study for a medical journal. This immediately felt like exactly what Max's eternal self was sharing. Even though this was a challenging experience for both of them, it was meant to be. The success of his treatment and recovery would help other animals, and his soul's purpose was to be this case study for a groundbreaking cancer treatment.

Monica and Max went on their road trip and had an amazing time in the Grand Canyon. I'm sure you will be glad to know that Max also fully recovered and lived to be 15 years old. He will forever be honored and remembered as the first dog to respond well to a new type of cancer treatment.

In the next chapter, we are going deeper into how you can create your own database of communication signs, symbols, messages, and feelings to better understand and interpret the information animals send you.

Journal Reflections

1. Have you ever experienced telepathic, energetic, or intuitive communication with an animal? Describe your experiences.
2. What are your intentions and goals as you embark on your animal communication journey?
3. How would you like to feel when successfully communicating with animals?
4. Are there any specific animals you would like to understand more deeply and communicate with more effectively? Who are they and what are some of your goals in learning with them?

ESTABLISHING YOUR INTERNAL DATABASE AND RITUAL

Good news: you already have an internal database of signs and symbols, which is simply your personal record of which things have certain meanings and inspire specific feelings. I'll prove it to you: What do you think of when you picture a big yellow bus? If you immediately thought of school and young kids with backpacks, that is because your internal database has learned to associate those things with one another. Throughout our lives as we learn and grow, everyone innately creates their own database of signs and symbols.

With consistent practice of using your intuition and applying it to communication with animals, you will begin to develop an internal database specifically for this purpose. In addition to supporting animal communication, your database helps you deepen your connection to the spiritual realm and gain greater insight and guidance on your path. It also serves as a powerful tool that builds your confidence to trust not only your intuitive guidance, but also the guidance from the Universe and nature that is always available all around you. By paying attention, recording your experiences, and reflecting on their meaning, you can cultivate a deeper awareness of yourself and the messages you receive.

Recognizing and Interpreting Symbols

Over time, your internal database becomes a rich and complex web of associations and meanings that inform your understanding of the world. Some things have a universal meaning while others may be more nuanced and specific to you, such as a song, smell, or object that holds significant meaning based on an experience or memory. Let's look at a few examples.

A stop sign is a symbol we are all familiar with; it is already stored within your internal database. So when there is a message related to stopping or pausing, perhaps this symbol that represents "stop" will pop up in your mind. Now imagine you are intuitively communicating with a cat, and you ask them a question about potentially adopting another cat. As you tune in to them, you suddenly see a stop sign. This could be the cat telling you that they do not like the idea; the stop sign is literally them saying *no, stop.*

I often also use the example of an apple. I might be communicating with an animal and see an apple. For me, an apple has a few potential meanings. It might symbolize the gift a child offers to a teacher in class to thank them. This could mean the animal is grateful for some form of teaching or training they have received and is acknowledging it. This could be more literal—the animal sharing that they like to eat apples. Or, maybe there is a person around them who is wearing pajamas with an apple print and the animal is simply calling attention to something they see as a way of validating our intuitive communication. The apple could even be related to something that resembles the fruit, such as a doorstop people are always tripping over.

You see, animal communication is more of an art than a science! There are a few important points to keep in mind, especially as you are building up your database:

1. *The signs you receive are representations of your mind's interpretation.* When we communicate with humans, it feels more straightforward because we can use the same language where words have the same meaning for both of us. But in my experience, animals

communicate with humans through energetic vibrations—not words or visual images. Then, my intuition interprets that energy to best articulate the message it contains. The energetic communication processes through my internal database, and I am served with an image in my mind's eye to help me interpret and understand the message. I call this the *energetic vocabulary.*

2. *Openness, flexibility, and patience are key.* I always tell clients that I will share whatever signs, symbols, and messages come through, even if they seem silly or unrelated. They may seem random to me, but they might mean everything to the person. For example, I was once communicating with a dog and kept seeing him beating on a drum like someone in a marching band. This seemed so strange until I learned that the dog's favorite toy was a mouse holding a drum. Along with this, don't be afraid to get curious. If you aren't quite understanding the signs you're getting, sit with them and delve deeper. It can be easy to immediately discount or ignore what doesn't make sense, but if you can lean in, what first seems confusing can open up with a deeper meaning.
3. *Animals perceive the world differently than humans.* This means that sometimes they will share a sign that we view as being or meaning one thing, but it's something different to the animal. Such as if I might be connecting with an animal and see a cookie in my mind's eye. Maybe the animal is saying they love cookies, or it could be their person calls them "Cookie" as a nickname. Animal communication can be like a game of charades, where you are receiving telepathic images that aren't obvious at first. Sometimes you have to do a little detective work to uncover the meaning.

Here are a couple of stories that help illustrate these important points. Early on in my professional career as an animal communicator, there were occasions when I would see a symbol or image, or even hear a song in my head, and think to myself, *Don't bother sharing that. It won't be related and the client will question my skills—or my sanity.* A specific example comes to mind when I kept hearing the song "You Are My Sunshine" in my head while communicating with a dog. But I didn't share it with the client because I thought she'd think I was being silly or cliché. Several minutes later, the woman shared with me that she sings that very song to her dog every night before bed. I quickly responded with, "I was hearing that in my head!" That certainly taught me to not hold back, to just go with the flow and share what comes to me. If it doesn't resonate with the person, so be it. I would rather share something not related than miss out on a message that could hold deep meaning for them.

There was another time when a client, Cynthia, asked me to communicate with her cat, Deva, because she was curious about her likes and dislikes. When I was connecting with Deva, I kept seeing what appeared like a ceiling fan, and I felt that Deva loved to watch it and keep an eye on it. This was not ringing a bell for Cynthia, but I suggested she keep it in mind in case something made sense after our time together. A few days later, she e-mailed with excitement. Cynthia owned a thrift store and there was a turnstile at the front door that resembled a fan in its shape and movement. Deva would sit all day on top of the turnstile, greeting people as they entered and enjoying the clicking sound it made. This is exactly what Deva was showing me, and it's an example of how it is important to be flexible in our interpretation. As soon as I read Cynthia's message, everything in my mind's eye lit up, almost like a strobe light. When everything becomes bright like that or if I see fireworks in my mind's eye, that is always a validation and confirmation of a message.

Aspects to Consider

Different signs have different meanings for different people. While this is very personal, and you will grow to have your own interpretations of the symbols you see, I want to share a few other symbols that come up more frequently for me as well as the kinds of things I look out for in my sessions.

Animals often represent powerful messages. In many spiritual and cultural traditions, certain animals are seen as messengers or symbols representing specific qualities or insights. The idea is that the natural world and the spiritual world are closely connected, and messages from the divine, spirit, or universal consciousness can come through in the form of animal encounters. These animals, including winged creatures like owls, crows, cardinals, and butterflies, are often seen as particularly symbolic.

Owls are associated with wisdom, knowledge, and intuition in many cultures. An encounter with an owl may be interpreted as a message to trust your instincts. Crows are commonly linked to change due to their association with death and the afterlife in various cultures. Seeing a crow might suggest a forthcoming change or transition, or it could be a reminder to release old habits and beliefs that no longer serve you. Cardinals are often associated with renewed vitality and recognition of self-importance. They are known for their bright red color, which is seen as a symbol of vitality and faith. A cardinal may be a reminder to stay true to who you are. Butterflies are universal symbols of transformation due to their metamorphosis. From a spiritual perspective, seeing a butterfly could be a message to embrace changes in your life and navigate them with grace and lightness.

Other natural objects also frequently come up. A flower opening can represent positive growth, expansion, and forward movement. If the flower is a red rose, I may associate the red with a very grounded energy, infused with a lot of love. If the flower is wilting, this may mean the animal needs care in some form, that they are literally dehydrated, or they need to be fed in some way, whether mentally, emotionally, or physically. A figure eight or infinity symbol represents a deep, eternal soulmate connection between two individuals, such as between a person and an animal or between two animals.

Colors, sounds, and scents all may hold specific meanings. For me, light blue represents communication and sharing one's voice. So if I am connecting with an animal and I see this shade of blue, I will pause to consider how expression and communication applies to the animal. As I tune in deeper, more energy may come through in the communication to fill in the gaps and clarify a point or message. I may interpret that the animal loves to use their voice by barking or meowing, or that they have a lot to share and are looking for ways to express themselves more.

Growing Your Internal Database

The first step in cultivating an internal database is to pay closer attention to what is happening around you. This is where the present moment mindfulness that you've been practicing comes in. Be aware of the events, encounters, and experiences that you have in your daily life—not just those connected with your animal communication. Take note of patterns, symbols, or messages that show up repeatedly. These could be signs you receive as intuitive downloads in your dreams, or they could be actual symbols around you in advertisements, online, on the radio, on clothing, or even in nature.

For example, maybe you are driving and stop signs catch your attention more than usual, even when they don't apply to your direction in traffic. As you continue about your day, you keep seeing a stop sign in your mind's eye. This could be your intuition tuning in to the energetic vibrations around you and repeatedly showing you this sign to get you to slow down or pause. Or it could be an alert to not move forward with a decision you are considering.

As you cultivate your internal database, use your journal to record the signs and symbols that you encounter, how they make you feel, and what you associate with them. Make note of what jumps out at you, including your thoughts and feelings about the experience and any insights you receive. Even when the messages don't seem related to animals, add them to your journal. Then when you are intuitively communicating with animals, they will be within you to help articulate the energetic vibrations the animals are sharing with you. The more touchstones you have, the easier your communication will become.

In addition to simply recording what you experience, take time to reflect on your experiences and consider what they might mean. A good practice when first starting your database journaling is to look back at the data you collected at the end of each month. Look for patterns or themes in what you encountered. Consider how they relate to your spiritual path and personal growth as well as your communication with animals.

When interpreting signs and symbols, it is important to trust your intuition. Allow yourself to be guided by your inner wisdom rather than trying to force a particular interpretation. Once you have begun to develop a record of the signs and symbols you encounter, you can use it as a tool for guidance and insight. When faced with a question or challenge, refer back to your journal to see if there are any relevant symbols that may provide support.

Practice: Recognizing Signs

In the beginning, I recommend being specific and purposeful in your collecting of observations. As you grow in your practice, these things will become second nature, but while you are learning, doing them deliberately will lead to more consistent outcomes. Here is a step-by-step practice for recognizing signs, symbols, and messages and learning from them:

1. *Set an intention* to recognize signs, symbols, and messages that are meant to guide and support you on your path.
2. *Pay attention* throughout your day to the events, encounters, and experiences that you have. Look for patterns or repeating themes that may be significant.

3. ***Record your experiences*** in the moment by keeping your journal with you. Write down the date, location, and any other details that seem relevant. Include your thoughts and feelings about the experience.
4. ***Reflect on your experiences*** and consider what they might mean for you and what feelings you associate with them. Look for patterns or themes in the signs and symbols that you have encountered. Consider how they relate to your spiritual path, personal growth, and your animal communication energetic vocabulary.
5. ***Interpret the signs and symbols*** and trust your intuition. Allow yourself to be guided by your inner wisdom rather than trying to force a particular interpretation. Consider the context of the experience and any other history you have with those signs.
6. ***Take action*** after interpreting to integrate the message into your life as well as your animal communication.
7. ***Express gratitude*** for the signs and symbols that you receive. Whether through prayer, meditation, or simply saying thank you, take a moment to acknowledge any guidance.

Foundations for Successful Sessions

Before beginning a session, find a quiet and peaceful environment free of distractions and interruptions. Have your journal with you so you can write down any communication or information that comes up during the session. Each session can follow this series of steps:

1. *Ground and center* by taking a few minutes to meditate and tune in to what is yours and not yours through whatever method works best for you, such as a body scan or objective observer meditation. This helps you establish a sense of calm and focus within yourself.

2. *Set an intention* for calm and confident communication and what you hope to achieve. This might be to gain insight into the animal's behavior, address a specific concern, share messages of love, or have a conversation to learn more about them. Visualize your intention to communicate and ask for their permission and willingness to participate in the process. Approach the communication session with an open mind and heart.
3. *Connect with the animal* either through a photo or by observing them in your physical space. If you are communicating in person, you may sit with them however feels most comfortable. If they can't sit still, that is perfectly fine. If you are connecting remotely, have one or more photos of the animal in front of you. This could be on a computer or phone or printed out. Surrender to the moment and go with the flow, believing that you can connect with them regardless of where they are or what they are doing. Allow yourself as much time as you need, and don't rush. It takes as long as it takes. You might need 3 minutes or 20 minutes to get into a calm and centered space before communication opens.
4. *Sending:* Use your intuition to tune in to the animal's energy and communicate through thoughts, emotions, images, or sensations. We will go through various processes for communication in Chapter 7.
5. *Receiving:* Allow the animal to respond in their own way and trust the information that comes through. You may open and close your eyes however you feel guided throughout a session. This is where the four clairs come into play. Recognize *how* you are receiving information from the animal—signs, symbols, or visions in your mind's eye; audibly hearing sounds, messages, or voices; feeling or

sensing something deep within; or having an inner knowing or understanding of what you are receiving. Really listen with all your senses and being.

6. *Recording:* A good practice is to focus more on allowing yourself to be present in that energetic space with the animal. When we think intellectually, take ourselves too seriously, and feel pressured to write everything down, that can be distracting. But at the same time, if you feel there is great information coming through that you want to jot down so you don't forget it, go for it. With practice, you will create a system and flow that is best for you.
7. *Interpret and validate* the information you receive using your intuition, other tools or techniques, and through discussion with the animal or their caregiver to ensure accuracy.
8. *Express gratitude* for the opportunity to communicate and respect for the animal's energy and messages. Complete the session by clearing your energy through the use of light and gently cutting any cords between you and the animal (refer back to the practices in Chapter 4).

I recommend giving yourself at least a half hour for each session: 5 or so minutes of grounding and centering yourself, 10 to 20 minutes of communication, and then another 10 to 20 minutes to interpret and decipher the information you received. Take more time if you need to. Think of each session as simply an opportunity for you to practice, find your groove, and see what works best for you. If you feel no information came through from the animal, that's perfectly fine. Communication likely did come through, but you may have missed it or perhaps doubted it as one of your own thoughts or feelings. This is just something that you learn to recognize the more you practice. Let go of discouragement and know that when you surrender to the process and keep practicing, you get out of your own way in receiving and understanding the subtle pieces of information flowing to you and through you. Perhaps, too, you circle back to your notes from a previous communication session to see if any new information comes through upon review and reflection.

Establishing Your Ritual

Creating a ritual, or a consistent series of actions that form the framework of your sessions, can help to cultivate centeredness and focus, deepen your connection to spirit, and enhance your intuition and communication skills. The main point of a ritual is to create a dedicated sacred space and process with intention and meaning, so this can be very personal. Your sacred space can be any physical location that is dedicated to spiritual practice and connection. It can be indoors or outdoors and can take many forms, such as a meditation room, garden, shrine, small corner of a room, or any other location that holds personal or cultural significance.

When crafting your ritual, it is important to be true to your energy, appreciations, and values. Focus on what resonates with you and feels meaningful and powerful. When you turn animal communication into a ritual, it can make it easier to find the time for it each day. While not required for successful meditation and animal communication, a consistent ritual will help when you are starting. A sacred space will also help promote peace and deepen your connection to the spiritual realm. It can provide you with a refuge from the stresses of everyday life, allowing you to recharge and renew your energy.

Creating an Altar

One way to designate a sacred space is by creating an *altar*, which simply means decorating a specific spot with items that hold personal or spiritual significance as well as things that bring you joy and lift your spirits. Many different cultures and religions use altars, but for the purposes of this book, we will be referring more generally to a sacred space that you create for yourself to reflect, renew, and focus on your spiritual practice of choice. On your altar, you might include candles, crystals, iconic symbols, art, items related to your pets, or photos. I have several sacred spaces in my home and yard because I love to have areas that uplift my spirits. In my walk-in closet, my dresser is draped with beautiful purple fabric, and I display a few collars from my dogs who have passed along with some crystals and a statue of Michael the Archangel (whom I was named after) that my sister gifted me. In my workspace, I display a large amethyst on a stand, a Himalayan salt lamp, a beautiful orchid plant, and framed photos of my animals. I also have a sacred corner of my yard with a Buddha statue and waterfall fountain surrounded by lush bamboo. Even when I just walk by these spaces and look at them, I can feel my energetic vibration lift and a peacefulness wash over me. I also notice a shift in my animals' energy when they are in these sacred spaces. I have selected each item in the sacred space with intention, love, and joy, and I keep these spaces neat and clean. This energy is what makes the spaces powerful.

While all these spaces are meaningful to me, I mainly practice animal communication in my office. In the same way, I encourage you to pick one specific sacred space for your meditation and animal communication and make it a ritual of going to that space for your practice every day. A dedicated space trains your mind, body, and spirit to adjust to the necessary energetic vibrations each time you are within it. This is similar to how we are trained to fall asleep when we are in bed—we spend so much time sleeping there that the space holds a restful energy. Your consciousness recognizes the energy of the environment and adjusts for the activity at hand. In the same way, as you get into a routine with animal communication in your sacred space, it will become easier and faster to slip into the mindset.

Your altar doesn't have to be permanent if you don't have a great space to set one up. For example, I encouraged my client Melanie to create a sacred space, but she was a bit concerned her family and friends would think it was a little out there. So she created a mobile altar that she kept in a drawer. She would unfold a beautiful multicolored tapestry and set out a pink candle, a photo of her deceased grandmother, a photo of her horse from childhood, and some of her cat's toys. She could easily set the items up for her sessions, and then quickly tuck them away when finished. Plus, Melanie frequently traveled for work, so she would pack up her little altar into her suitcase and never miss a day.

Crafting an Invocation

The next step of your ritual is to create an invocation to begin and end your sessions. An *invocation* is a prayer or ritualistic sentence or two that is used to call upon Spirit, a higher power, divine energy, and your intuition. Beginning a session with an invocation can be a very powerful way to tune in to your intuition and the animal more formally. I like to think of it as turning a key in the ignition of a car to start the engine. With practice and repetition, our consciousness gets used to the intentional energy of the words and innately understands how to turn on our intuition and elevate our energetic vibration for clear communication. In my experience, an invocation is not required to have a successful session, but it definitely can help, especially as you strengthen your abilities. When your session is complete, reciting a few closing words will help to energetically end the session for you and the animal.

Here are examples of invocations:

- *Beginning Invocation:* "Great Spirit, Universe, Divine Creator, and All That Is, please support and guide me as I connect with this animal on an energetic level. Help me to open my mind, heart, and intuition to their energy and messages, and communicate with them in a way that is respectful and compassionate. May my communication be guided by the highest good for this animal and all beings. Thank you for this opportunity to connect with the divine energy that surrounds us all."

- *Ending Invocation:* "Great Spirit, Universe, Divine Creator, and All That Is, I offer my deepest gratitude for the opportunity to connect with this animal on an energetic level. Thank you for the messages and insights that have been shared and the guidance and support that has been given. I honor the energy and spirit of this animal and offer my love and blessings for their well-being. May our communication continue to be guided by the highest good for all beings. Thank you for this sacred connection. And so it is."

The phrase "and so it is" is often used as a closing affirmation or declaration in spiritual texts, prayers, or meditations. In essence, it is a way of saying that the statement or intention is accepted as reality or as a manifestation of divine truth.

You can use this as a starting place, but I encourage you to consider adapting it or creating your own based on the words, intentions, and energies that are most aligned with you and the animals you are communicating with. It could be one short sentence or several, whatever feels right for you and helps you in your communication.

In the next chapter, I will share techniques and processes for animal communication, and we will be putting into practice everything you have been learning. This is when you get on the bike and start riding, so to speak. Get ready to begin experiencing the energy of animals and yourself more deeply and profoundly!

Journal Reflections

1. What kinds of symbols and signs are already a part of your internal database? For example, what are your connotations with the color orange, the sound of the ocean, or the scent of watermelon? Take some time to journal about what the idea of the internal database means to you.

2. What symbols often come into your consciousness or dreams that represent thoughts, feelings, emotions, or experiences? Choose one or two that you are drawn to and track them closely. Keep journaling what symbols come to you and consider what kinds of meanings they could represent.
3. What kinds of sacred spaces do you already have in your house? Where can you create a sacred space specifically for animal communication? What items come to mind that you love, lift your spirits, and would provide supportive energy for you in this space?
4. Create an invocation for your animal communication sessions. Spend some time meditating or try freewriting (see page 8) to tune inward. Find the words and energy you feel are most supportive for you and try it out for a week. You can always adjust as needed!

TECHNIQUES FOR COMMUNICATION

Now that you have been meditating and strengthening your intuitive muscles, we are going to get into the fun part—practicing intuitive animal communication. In this chapter, I will share techniques that have worked well for me and others over the years. Everyone is different, so some may resonate more with you than others. Plus, as you practice, you will begin developing a unique method of communicating with animals. These techniques can be used as a springboard for connecting, and they will provide you with opportunities to further cultivate your skills.

After reading through this chapter, I suggest choosing one technique to practice five or more times. Any intuitive work is a process, so you don't want to give up after one or two tries if it seems to not be working. Even if you are not feeling results at first, think of it as planting seeds. Right after you plant them, it doesn't seem like anything is there. You need to regularly water the seeds and nurture them before they peek up through the soil. Practicing and not giving up on a technique is a way of watering your intuitive seeds. You may have a few sessions where you feel you are not successfully communicating with the animal, but then after the third or fourth try, the floodgates may open up. Whether or not you find a strong connection with any one technique, I also recommend giving each a try before deciding on one to stick with. Trust your intuition—if you are feeling it is time to try a different process, go for it. For now, I recommend using one process at a time, though we will discuss combining processes in Chapter 9.

With each technique, I encourage you to take five or so minutes to meditate before communicating to ground and center yourself. (You can use whichever technique you like, but see the Mindful Breathing practice, page 17; Grounding and Centering Meditation practice, page 14; and Objective Observer Meditation practice, page 73.) This way, you will have objective clarity of which thoughts, feelings, and emotions belong to you and which belong to the animal. Then grab your journal so you can write down anything that comes through, so you can refer back to it later. Practice and patience are key; in time, you will become more skilled at interpreting the messages and symbols that come to you.

Let's first walk through the basic steps on how to communicate remotely using a photo as well as in person. Either of these processes can be used for any of the techniques that follow.

Communicating Using a Photo

1. Grab your journal and find a peaceful location (such as your sacred space) where you can focus without any distractions.
2. Take a few deep breaths to calm your mind and body.
3. Recite your opening invocation.
4. Hold the photo in your hands or place it in front of you and focus your attention on the animal's energy and presence.
5. Allow yourself to connect with the animal by setting an intention to do so and using your intuition. Take note of any thoughts, feelings, or emotions that come to mind as you connect with the animal's energy.
6. Begin to communicate with the animal using one of the techniques that follow. Visualize and feel the energetic positivity and well-being that encapsulates your communication with the animal. Write down in your journal anything you feel guided to note.
7. Once you have finished the communication session, recite your closing invocation. Then take a few deep breaths and

express gratitude for the opportunity to connect with the animal in such a deep and intimate way. Take time after your session to contemplate and interpret your notes.

Communicating in Person

1. Grab your journal and find a comfortable location for both you and the animal where you can focus without any distractions. This could be your sacred space, but it doesn't need to be.
2. Take a few deep breaths to help calm your mind and body.
3. Recite your opening invocation.
4. Allow yourself to connect with the animal's energy and presence by setting an intention to do so and using your intuition. Take note of any thoughts, feelings, or emotions that come to mind as you connect with the animal's energy.
5. Observe the animal's body language, paying close attention to their movements and expressions. Use your intuition to feel any sensations or emotions that come to mind.
6. Begin to communicate with the animal using one of the techniques that follow. If possible, you may feel guided to touch or hold the animal gently and respectfully, using your intuition to sense any sensations or emotions. Write down in your journal anything you feel guided to note.
7. Continue to communicate with the animal as long as feels comfortable and natural, being respectful of their boundaries and needs. Once you have finished the communication session, recite your closing invocation. Then take a few deep breaths and express gratitude for the opportunity to connect with the animal in such a deep and intimate way. Take time after your session to contemplate and interpret your notes.

Heart-to-Heart Technique

When it comes to our perception, relationship, and understanding of animals, our love for them is usually what we lead with. Their beauty, special personalities, and close connection to nature inspire a caring and loving response within us. Because of this, energetically communicating from your heart center to theirs can be a quick way to profoundly connect.

This technique is especially helpful with sensitive animals or those who have experienced trauma and also when considering a potential adoptee. When I first met my dog Harlow, for example, I immediately connected with her this way and could feel her sensitive and loving energy. She wanted peace and simple happiness, a place to relax, and to feel the freedom to express the joyous, funny side of her personality that had not yet come out. I also felt her receptivity to a heart-to-heart connection because she truly had a big heart, so it was a powerful energetic space within her for clear communication.

1. Follow steps 1 through 5 of the Communicating Using a Photo or Communicating in Person (pages 112–113) instructions, depending on your situation.
2. Bring your focus to your heart center, in the middle of your chest. Visualize or feel your heart opening up and imagine a warm and loving light radiating from it.
3. Visualize a beautiful beam of light connecting your heart to the animal's heart. Imagine the energy flowing freely between you both. See and feel this beam of light taking the form of a figure eight or infinity symbol, perhaps moving or pulsing with your energy exchange. Allow yourself to feel the flow back and forth through this symbol. You may also imagine you are both surrounded by a warm and loving light as you communicate together.
4. Feel yourself sending and receiving messages and communication through this heart-to-heart beam of light and energetic connection. Use visualization and intention to promote understanding. Take your time as you communicate

with the animal, allowing the process to unfold naturally and intuitively.

5. Finish with step 7 of the Communicating Using a Photo or Communicating in Person (pages 112–113) instructions, depending on your situation.

Third Eye Connection

The third eye, also known as the "inner eye" or *ajna* in Sanskrit, is a powerful concept that originates from ancient Indian spiritual traditions. It's often associated with the sixth chakra, located in the middle of the forehead, slightly above the space between the eyebrows. Traditionally, the third eye is viewed as a symbol of enlightenment and inner wisdom. It's believed to be the center of insight, where one can achieve a higher state of consciousness and perceive the subtler realities of existence. This metaphysical eye isn't about physical sight, but about intuition, awareness, and spiritual knowledge. The power of the third eye lies in its ability to transcend the physical and material world, enabling a deeper understanding of one's inner self and the Universe. It's associated with psychic abilities such as clairvoyance and precognition.

Through awakening or opening the third eye, we can tap into an intuitive sensibility and inner wisdom that lies beyond ordinary perception. This may enhance one's ability to perceive energies, auras, or other nonphysical phenomena.

The Third Eye Connection is a productive way to communicate because it helps narrow focus to a specific location. If you are sighted, you are naturally conditioned to understand the world through your eyes, so it can be instinctual to tune in through this area when your eyes are closed. Plus, the third eye area is a powerful energy center and tuning in to our intuition through that specific spot strengthens our abilities. I often use the Third Eye Connection with my brilliant Australian shepherd, Atticus. In addition to training, I would also regularly connect with him using this technique. When he was a puppy, he would overcorrect other dogs when he felt their energy was too high. I felt him sharing that he was just trying to help create order and calm things down. I communicated that it was not his job to correct other dogs and shared visions and feelings of him being more passive and understanding, turning to me for guidance and leadership when he felt others needed to be dealt with. Through this connection, we were able to better communicate with each other and Atticus learned how to make better choices.

1. Follow steps 1 through 5 of the Communicating Using a Photo or Communicating in Person (pages 112–113) instructions, depending on your situation.
2. As you continue to breathe slowly and deeply, bring your attention to your third eye center.
3. Imagine your third eye gently filling with light and visualize a beam of light extending out toward the animal, connecting your third eye to the animal's third eye. Feel the energy flowing freely between you and the animal.
4. Use your intuition to connect with the animal's energy and presence. Focus your attention on any messages or insights that may come through. As you feel guided, imagine that you are sharing messages with the animal through this connection of light. Focus your intention to amplify understanding and connection. Take your time as you communicate, allowing the process to unfold naturally and intuitively.
5. Finish with step 7 of the Communicating Using a Photo or Communicating in Person (pages 112–113) instructions, depending on your situation.

Screen and Flower Technique

I use this process when I am gathering (rather than sharing) information. This technique involves tuning in to your intuition through the third eye and imagining a screen there, within your mind's eye, similar to a computer monitor or television screen. Visualize a flower appearing on the screen and watch to see what happens as you ask questions. The flower may transform into a symbol or another object of some kind. It may change shape and fill with colors. Or maybe it blooms or wilts.

One time a client asked me how their bunny felt about their dog sibling. Using this technique, the flower started trembling, lost all color, and became frozen like ice. Then I saw a bowling ball roll fast across the screen and shatter the frozen flower in its path. The woman told me that made sense because the bunny would always start shaking and freeze up in the presence of the dog. The bunny was terrified, but the dog loved the bunny and would approach him with a lot of energy, charging at him with as much momentum as a bowling ball.

The flower on the screen serves as an oracle for you to simply observe. I think of this type of communication as an energy interpreter, where the communication from the animal flows through the screen and the flower offers a visual interpretation of the information for you to understand.

1. Follow steps 1 through 5 of the Communicating Using a Photo or Communicating in Person (pages 112–113) instructions, depending on your situation.
2. As you continue to breathe slowly and deeply, bring your attention to your third eye center.
3. Visualize there is a screen in the center of your mind's eye with a grounding cord that extends down and connects to the core of the earth, keeping you clear and focused.
4. See the outline of a flower appear on your screen. I like to imagine it as the outline of a rose, almost like one in a coloring book ready to be filled with color.

5. Imagine you are writing out or typing the animal's name on the screen above the flower and drop the name down into the flower. Relax and observe what happens to the flower without judgment.
6. Allow yourself to connect with the animal's energy and presence by using your intuition as you continue to observe the flower.
7. Imagine you are writing or typing out each question on the screen above the flower, and then drop it into the flower. Be patient and allow the flower to show you the answers. Use your imagination and be open to whatever is revealed to you, even—and ***especially***—if it seems random or unrelated.
8. Finish with step 7 of the Communicating Using a Photo or Communicating in Person (pages 112–113) instructions, depending on your situation.

Golden Circle Technique

The feeling of being within a special circle provides safety and can also help block out distractions for both you and the animal. I once was communicating with a cat that was riddled with anxiety because it had once been attacked on the street by another cat. It was hard to keep her focused in a communication session. After trying to connect with her a few different ways, I found that the Golden Circle was the easiest and most supportive for her because it offered her a safe space to slip into and take a break from her chronic anxiety.

In the Golden Circle I created for her, I was able to get 100 percent of her attention and energy. Because she spent time communicating with me in the Golden Circle over a series of sessions, I was able to calm and soothe her. Over time, she became less nervous and anxious, eventually letting go of her PTSD from the fight.

1. Follow steps 1 through 5 of the Communicating Using a Photo or Communicating in Person (pages 112–113) instructions, depending on your situation.
2. Imagine yourself within a golden circle on the ground. This circle provides a safe, protective, and loving space filled with warm and positive energy. Feel its sacred power, however that presents to you.
3. Use your intuition and intention to connect with the animal's energy and invite them to join you in the sacred golden circle. You may see them entering the circle or easily appearing within it.
4. As you begin to communicate and connect, imagine the golden circle begins to glow brightly, blocking out the outer world and keeping you both safe. Visualize you are having a telepathic and energetic conversation where you can ask questions and listen to each other.
5. As your communication draws to a close, the light cast by the circle will start to dim. See the animal slowly leaving the circle and when you are ready, you may also step out of it. Finish with step 7 of the Communicating Using a Photo or Communicating in Person (pages 112–113) instructions, depending on your situation.

Vibrational Match Connection

As we have discussed, everything in the Universe vibrates at a certain frequency, and by matching our vibration with that of an animal, we can tune in to them at a deeper level. A great way to go about this is to imagine you are tuning a radio dial until it picks up a station. You may even imagine yourself as the radio frequency, energetically vibrating like a wavy line. As you begin to tune in to the animal, you see them as a different wavy line, maybe with higher or wider peaks and crests. Once aware of where you each are, you can gently alter your wave so it aligns with theirs, allowing you to blend and energetically vibrate at the same level.

This process can be especially helpful when communicating with animals in the afterlife and lost animals—both of which we will explore further in Chapters 8 and 9.

1. Follow steps 1 through 5 of the Communicating Using a Photo or Communicating in Person (pages 112–113) instructions, depending on your situation.
2. Begin to visualize yourself as a vibrational match to the animal, adjusting to meet their frequency level. Imagine your energy aligning with the animal's energy, becoming one and the same.
3. Allow yourself to feel any emotions or sensations that arise as you become a vibrational match. Use your intuition to connect with the animal, allowing the connection to deepen.
4. Use visualization and intention to promote understanding and connection in conversation with the animal as you ask questions and listen. Imagine the energy flowing freely between you and the animal.
5. Finish with step 7 of the Communicating Using a Photo or Communicating in Person (pages 112–113) instructions, depending on your situation.

Great Glass Elevator Technique

When I was around 10 years old, one of my favorite books was ***Charlie and the Great Glass Elevator*** by Roald Dahl, the sequel to ***Charlie and the Chocolate Factory***. I was mesmerized by the story of Charlie traveling through space and time in a strange glass elevator, partly because there was a beautiful, state-of-the-art, all-glass elevator in Water Tower Place, a mall in Chicago where we often went shopping. To me, it almost resembled a large crystal, and getting to ride in it was an exciting treat.

At the time, I was having recurring dreams of getting on that glass elevator and choosing a special floor number. When I stepped out, my dog Corky was there to meet me. We would have conversations, and, because he was a challenging dog, I would ask him questions about his behavior. Those dreams did offer insights into his actions, and I would wake feeling as though they were real. Because this was such a vivid experience, riding the glass elevator became one of my first techniques for intuitively communicating with animals. The whole act of imagining you are stepping into a container that rises and delivers you to a sacred and intimate meeting makes it easier to tune in to a special intuitive space.

1. Follow steps 1 through 5 of the Communicating Using a Photo or Communicating in Person (pages 112–113) instructions, depending on your situation.
2. In your mind's eye, begin to visualize a great glass elevator in front of you. Imagine yourself stepping into the elevator. There are buttons with floor numbers inside next to the door. Allow your intuition to guide you to choose a floor number and see that button light up.
3. The elevator rises to your chosen floor, and the doors open. Step out of the elevator and take a look around. This may be a wide open and glowing space. Or you may visualize it with specific details. Settle into the space and call upon the animal to join you here. You may see them enter the space from the left, right, or center. Or they may simply materialize in front of you.
4. Greet the animal and begin to communicate with them using your intuition, asking questions or simply allowing yourself to receive any messages or insights that come to you.
5. When you are ready to end the communication session, thank the animal for your time together and see them exiting the space however they choose to do so. Then step back into the great glass elevator and select the button for the ground floor. As the elevator descends, release any remaining energy or emotions.

6. Once you have returned to the bottom, see the doors open and exit the elevator. Finish with step 7 of the Communicating Using a Photo or Communicating in Person (pages 112–113) instructions, depending on your situation.

Sacred Temple Technique

Temples are spiritual spaces where individuals can connect with the divine, seek guidance, and find solace. They also serve as a physical manifestation of safety and tranquility. Because of the powerful symbolism associated with temples, creating one to be your meeting place for animal communication can support a clear connection. Similar to having a sacred space in your home, having a sacred temple within your consciousness quickly shifts your consciousness into a realm of intuitive clarity. Each time you visit the sacred temple, your energy more quickly adjusts to the frequency for communication. Plus, it can be fun to be imaginative in creating such a special place. The animals you communicate with will respond positively to the beautiful energy of this temple too.

1. Follow steps 1 through 5 of the Communicating Using a Photo or Communicating in Person (pages 112–113) instructions, depending on your situation.
2. Imagine you are standing on the outside edge of a beautiful forest. The temperature is perfect, and the sun is gently shining. You're surrounded by the harmonic sounds of nature. There is a pathway that you follow through all the vegetation. You see a glowing sacred temple with a few steps up to a platform. You walk up the steps and see that there are no walls—the temple is open to the natural environment but has a roof supported by four pillars. Move into the center where, looking up, you find a circular opening in the roof that allows natural light to shine down upon you.

3. Invite the animal to join you in this sacred temple and visualize them entering from the opposite end, facing you. Feel the animal meeting you in the center, under the light. Focus your attention on the animal's energy and presence.
4. Use visualization and intention to amplify understanding and connection with the animal. Imagine the energy flowing freely between you both. Take your time as you communicate with the animal, allowing the process to unfold naturally and intuitively.
5. Once you have finished the communication session, thank the animal and watch them exit. When you are ready, leave the way you entered. Follow the pathway through the lush forest to the outer edge where you began. Finish with step 7 of the Communicating Using a Photo or Communicating in Person (pages 112-113) instructions, depending on your situation.

Body Scanning Technique

By now, you'll be familiar with doing a body scan on yourself (see page 43). This technique applies that same concept to the animal you are working with. By intuitively scanning an animal's body, you can identify areas of tension or discomfort that may need attention. I always recommend checking with your veterinarian for further examination, diagnosis, and recommendations.

When I am connecting with an animal, much like a doctor examining a patient, I intuitively check in and ask questions about how they feel while scanning them. But sometimes, animals block out pain and physical challenges and respond that everything is fine. Remember that they are instinctual beings that naturally cope with what they are facing and mask their discomfort. For example, I had a client who asked me to communicate with their dog who had a lump under their right front limb. When I connected with the dog and asked how she felt about this lump, she responded, "I'm fine! I'm good! Let's get going, I've got things to do!" But as I scanned her body, I could feel a lot of heat and friction rubbing internally on the surrounding areas. It lit up in my mind's eye like a fireball. I recommended they see a vet to take a deeper look, and it turned out to be a precancerous growth that was easy to remove because they caught it early.

Sometimes when I am body scanning an animal and asking them how they are feeling, I will feel a part of my body mirroring their experience. For example, I may feel a sourness in my stomach or a tightness in my leg that is their experience, not my own. It's not that I am taking on their feelings and experience. Rather, I am tuning in to them and experiencing an example in my own body that gives me a reference of what they are feeling. This is why it is important to ground yourself and be clear on what your feelings are before scanning an animal.

1. Follow steps 1 through 5 of the Communicating Using a Photo or Communicating in Person (pages 112–113) instructions, depending on your situation.
2. Focus your attention on the animal's energy and presence. Begin to visualize a beam of light emanating from your third eye and scanning the animal's body. Imagine this beam of light moving over the animal's entire body, from the top of their head to the tips of their feet. You may imagine you are viewing their body from the side, and scanning them from one end to another, almost like you are running an imaginary x-ray over them. Or you may imagine them facing you and running the scan from the top of their head down to the bottom of their body, like a light shining down on them.
3. As you scan, use your intuition to feel any areas of tension or discomfort. Pay attention to any sensations, emotions, or images that come to mind. Things may appear differently for you, but if there is an area in question for me, it will often light up like a concentrated point of light. I will observe what

that spot is doing, such as changing shape or color, and note any other symbolisms that come through. When it's something that may be concerning, it usually will appear for me as reddish and warm, almost like it is heated or even on fire. I will also often see and feel in my mind's eye the flow of life-force energy moving through the body. If it feels like there are spots where energy is stalled or blocked, tune in deeper there.

4. Take your time as you communicate with the animal through the body scanning process, allowing it to unfold naturally and intuitively.
5. Finish with step 7 of the Communicating Using a Photo or Communicating in Person (pages 112–113) instructions, depending on your situation.

Automatic Writing Technique

Automatic writing is a form of channeling messages through writing on paper without control, allowing intuition to take over and guide the process. Automatic writing is typically done by holding a pen or pencil and allowing it to move across the paper without conscious thought or direction. The idea behind this is to bypass the conscious mind and tap into the subconscious mind and spirit, allowing for a deeper connection to intuition, creativity, and spiritual guidance. This can lead to unexpected insights that may not have been accessible through conscious thought alone.

A client of mine named Alyssa was an equestrian and the proud guardian of a beautiful horse named Bella. Bella had always been a strong, vibrant, and enthusiastic jumper, but recently, her performance had faltered, and she seemed less enthusiastic about her training sessions. Alyssa couldn't find any physical ailments, and her vet assured her that Bella was in excellent health. Perplexed, Alyssa decided to turn to intuitive animal communication using automatic writing to understand

Bella's sudden change in behavior. Alyssa chose a calm and quiet evening for her session, after Bella's dinner, when they both could relax. With a notebook and a pen in hand, Alyssa sat near Bella's stall, feeling her calming presence. After grounding herself through a short meditation, Alyssa set a clear intention to understand Bella's feelings and discover what was causing her behavior change. Without any conscious thought, she allowed her hand to move freely across the paper. As her pen flowed, words and phrases came forth: "Lights too bright. Noise. Clapping scares. Miss quiet rides. Miss our alone time."

Alyssa let her hand write without any judgment or interruption. After about 20 minutes, she slowly came out of her writing trance and started reviewing. She realized that Bella was feeling overwhelmed by the increased intensity of their training sessions and the lights, loud noises, and audience clapping during their practice shows. Inspired by these revelations, Alyssa modified Bella's training schedule, incorporating more quiet rides, just like old times. She also requested the training arena to adjust the lighting and noise levels during Bella's sessions. Over the next few weeks, Bella's performance improved, and she seemed happier and more enthusiastic. This experience reinforced Alyssa's belief in intuitive animal communication, giving her a valuable tool to understand her beloved horse on a deeper level. It reminded her that every animal, like humans, has its preferences and insecurities, and by respecting and addressing them, we can enhance our bond with them.

1. Prepare your space with a pen or pencil and plenty of blank sheets of paper or your journal. Follow steps 1 through 5 of the Communicating Using a Photo or Communicating in Person (pages 112–113) instructions, depending on your situation.
2. Set an intention to receive messages from the animal through automatic writing.
3. Begin to write down whatever comes to mind, without censoring or judging your thoughts. Allow yourself to write

freely, letting the words flow through you without conscious thought or control.

4. Use your intuition to sense the animal's energy and presence as you write, allowing yourself to receive any messages or insights that come to mind. Continue writing until you feel you have received all the messages and insights the animal has to offer.
5. Finish with step 7 of the Communicating Using a Photo or Communicating in Person (pages 112–113) instructions, depending on your situation.

Journal Reflections

1. Do you feel confident trying communication with an animal in person as well as remotely with a photo? If not, which feels more challenging and why?
2. After reading through this chapter, and before diving in, which techniques are you drawn to? Why do you think that is?
3. How do meditation and the exercises discussed in previous chapters support the techniques described here?
4. What are you most excited about as you begin to try these techniques? What are you most anxious about?

ANIMALS IN THE AFTERLIFE, SPIRIT ANIMALS, AND POWER ANIMALS

We can also communicate with animals who are deceased and have passed on to the spirit world or afterlife, though we can also think of it as a type of heaven. When someone passes away, their energy and consciousness transition to a different vibrational frequency, allowing them to exist in the realm of the spirit world. The higher their vibrational frequency, the more spiritually evolved and enlightened that being is. Therefore, spirits in the afterlife who have shed their physical body and its limitations can vibrate at a higher frequency and access higher levels of consciousness than living animals and humans. Before getting into how to connect with animals in the afterlife, I want to share what I have learned about what animals experience on the other side.

How Animals Experience the Afterlife

Like humans, animals have a team of angels, guides, and light beings who are with them throughout their lives. When it comes time for them to cross over, I feel that team prepare the animal's vibration for the afterlife. Sometimes I will feel their team with them for a year or so before the animal's death. As the time gets closer for the animal to pass, their team swoops in and surrounds them to help soothe and support. This team also offers the animal a taste of the afterlife, almost like dipping their toes into water before fully submerging in it. This helps the animal find peace with their life and acclimates their vibration for the frequency of the afterlife.

Animals on the other side always share feelings of amazing love, joy, exhilaration, excitement, and peace. These elevated positive feelings are off the charts, often beyond words, and because we are still on the earthly plane, we can only feel what they share to a certain extent.

Whenever I am connecting with an animal on the other side, I start to feel my vibration elevating to meet them where they are at. I feel a faster, elevated energy; sometimes my heart even beats faster, like the excitement of experiencing true happiness and fulfillment. The feelings of animals in the afterlife are different from living animals in that there is this deeper feeling of serenity along with elation and freedom because they are no longer confined to their physical body. This is often even more dominant when the animal had been struggling physically before crossing over, whether due to health issues or just being older. They almost always feel youthful, like they are "back to their old selves" and then some! I often experience their energy almost like a genie being let out of a bottle: once they are free of their physical body (the bottle), their essence can expand everywhere.

An animal's version of heaven is often doing what they loved to do with their human in life, such as a dog experiencing the thrill of being out on a sailboat for endless hours or a cat who loved sleeping in a beam of sunlight, relaxing on a glowing golden throne. I have felt a retired racing horse running once again and leading the pack like a streak of lightning with the energy of a huge smile on her face. I see and feel them in beautiful natural environments, galivanting as far as the eye can see with the sun shining down upon them.

In addition, I often feel that animals who have crossed over take advantage of opportunities to do things they were never able to do in life. I have seen dogs run, jump, roll around, and then climb up to the top of a tree and wave, saying, "Look what I can do now!" I have watched some animals unzip the outside of their body and drop it to the ground like a costume in a cartoon. Stepping out of it, they take a deep breath and celebrate the release of their physical self. I also often see them solely as a point of light, zipping around like a comet.

I also usually feel their presence all around their human loved ones. I have never felt an animal having regret, sadness, or any sort of resentment toward their human for anything that happened throughout their lives. Clients often ask me if their animal is okay in the afterlife or if the animal is mad at them for choices they feel guilty about, such as choosing euthanasia. My experience is that the energetic vibrations of the afterlife are so high and loving that these types of concerns simply do not exist for animals. I always feel that animals are happy to be in the afterlife, and they want to share that with their humans. I often feel them wanting to help their humans to let go of shame or regret. Though it is heartbreaking to lose a pet, the bonds of love never die; they merely change form.

The exciting part is that when an animal is on the other side, they are ready for the next chapter in their relationship with us. In spirit, they can guide, protect, direct, and support us, and they can truly be with us no matter where we are or what we do. Animals have also shown me that they can be in many places at once. They can experience their version of heaven while also being with various family members, no matter where the people are located throughout the world. I also feel them communicating with the other living animals with whom they shared their home.

In my experience, some animals get acclimated to the afterlife very quickly, and others experience a longer metamorphosis. Time does not really exist in the afterlife, but from our human perspective, it just takes as long as it takes. Some animals relax into the peace and freedom of it, and others immediately assume a proactive role as a guardian angel over their loved ones. Some will spend some time as a guardian angel and then elevate to higher energetic vibrations while still maintaining a connection with us. Many will be very present around their humans for a while to help them through grief and loss, making their presence known through dreams, signs, and feelings. Eventually, they don't come through as often because their supportive role is complete, and they are ready for the next phase in their soul's journey. That may include them reincarnating as another animal in our lives or the lives of people we do not know. I often feel that an animal who has crossed over arranges for a new animal to come into the life of their human. And the animal who is on the other side is excited and happy to be overseeing the new animal, mentoring them in their new relationship.

Soul Groups

I also experience many animals as part of a *soul group*. This is the idea that an animal's soul is connected to a larger group of other animal's souls (in addition to human souls) with whom they share a special bond and energetic connection. The purpose of soul groups is to offer support for spiritual growth and evolution. Sometimes, though not always, animals in a soul group experience life together. People often ask me if their new pet is a reincarnation of a previous pet or if their deceased pet will return to them in another form. What I see more commonly is that the animals are separate beings from the same soul group.

For example, Dennis asked me if his new puppy, Rufus, was reincarnated from his previous dog, Dobie, who had passed away. Dennis felt there were too many uncanny similarities between his two dogs—who had never known each other in life—for this not to be the case. I felt Dobie in spirit and recognized that he was the one who brought Rufus to Dennis. Dobie and Rufus were also of the same soul group, so it made sense that there were similarities in their personalities. Consider the analogy of a raindrop high in the sky falling into the atmosphere. As it falls, it splits into separate droplets that have similar qualities. But being from the same original drop, they retain a state of oneness. The original drop is the soul group. Dobie and Rufus are of the same soul group, so there are similarities, but just like with siblings, there are also some differences.

My main point here is that our beloved animals on the other side are always around and available to us. All we have to do is ask for a sign and be open to receiving it. As time does not exist on the other side, it's important to be patient, let go of expectations, and pay attention to everything around you. Spirit often speaks in subtle ways, and if we are distracted, we may miss the message. Let me introduce you to Dasher for an example of how this can work.

Dasher was a beloved dog of mine who passed away very suddenly from an autoimmune disorder when she was eight years old. It all happened very quickly and was a huge shock. Though I was terribly sad, I understood the blessing of her passing quickly, free from suffering, and going out on a high note, still youthful and very puppylike. The day after Dasher passed, I asked her for a sign of her presence on the other side. I suggested a red dragonfly, knowing that while dragonflies are native to my area, I had never seen one in my backyard in the many years we had lived in our home. I'd seen butterflies and other beautiful insects but never a red dragonfly.

The next day, sure enough, two red dragonflies appeared right outside my back door, dancing and chasing each other. One of them started flying into the house and swooping around as if it was trying to land on me! I went outside and sat on the edge of our fountain pond, and one of them landed right beside me on a blade of a plant, looking like it was smiling. I could feel Dasher's presence there, and I sat with that dragonfly for 20 minutes, just being present and grateful—grateful for the message from Dasher, grateful for the love, and grateful for the power of spirit that communicates with us in miraculous ways all the time.

The dragonfly came back every single day for exactly seven days, flying into the house to greet me and then sitting on the same blade of the plant near the fountain for hours each day. Each day I went out and sat with it for a bit, again feeling so grateful for my eternal connection with Dasher and the beauty and peacefulness of the moment. After the seventh day, the dragonfly no longer visited. But I still feel Dasher with me in so many other ways, surprising me and reminding me of her love and connection.

When I took Dasher to the emergency room before she passed, we were waiting in our car in the parking lot and out of nowhere, I started singing the Suzanne Vega song "Luka." I pulled it up on my phone and played it, singing along to Dasher while we waited. It is an upbeat-sounding song, even though the content is pretty heavy. Still, it helped lighten the somber mood.

Not long after Dasher's passing, I felt guided to a new puppy who appeared in our lives through a series of synchronicities. As soon as I saw him, I knew he was the one, the same way I have felt with all our other animals. I asked Dasher in spirit for guidance on a name and within minutes I heard the song again in my head, "My name is Luka. . . ." I was a bit shocked because I had completely forgotten about randomly playing and singing that song to Dasher, and I realized it was the last song she heard me sing to her.

I knew right away that was the name and thanked Dasher for her guidance. Sure enough, the puppy responded to the name immediately and licked my nose when I called him Luca (I decided to spell it with a *c*) for the first time. Interestingly, he shares many characteristics with Dasher, such as similar facial expressions and a very cute tendency to roll over on his back to lure me to the ground for love and pets. I can feel that they are of the same soul group, and I am appreciative of the familiarities between the two of them and the frequent reminders from Dasher that she is here with us, watching over us. At the same time, I have been able to develop a completely new and wonderfully rewarding relationship with Luca that is different from my relationship with Dasher.

Connecting with Animals Who Have Passed

Asking for a Sign

You don't have to be in deep meditation to feel animals in the afterlife. You can also experience them by paying close attention to the signs and messages—whether subtle or more obvious—that they send you from the other side. This could include seeing a particular creature repeatedly, such as when I saw the dragonflies, or smelling them or having a sense of their presence around you in certain situations.

If you are hoping to receive a sign from an animal who has passed, I recommend asking for one. It would be fine to ask for this sign in your mind or out loud or even to write about it in your journal, but what is most important is that you deeply feel what you are saying. When it comes to what to ask for, the more specific, the better! For example, rather than asking for them to send you a bird, ask for a specific color or species of bird. Then the key is to be open to how it comes to you. It may literally be a living bird that shows up on your windowsill, but it could also appear in an advertisement you see while flipping through a magazine or in a movie you are watching. When you receive your requested sign, spend some time acknowledging it in joyful contemplation, thanking your animal.

Julie was skeptical of connecting with Yogi, who recently passed away. She and her boyfriend felt sad and lost without their beloved cat, though, so they scheduled a session with me. I felt Yogi as an excited presence, happy to be guiding them from the other side. Yogi showed me many signs for them to look out for, including penguins, which seemed a little wild to Julie and her partner at first. But I kept seeing a penguin and that Yogi would be sending her a penguin confirming she was with them. I told Julie that the sign could come through in a lot of different forms: it could be on social media, or she could even be in the grocery store checkout aisle and there might be a stand filled with penguin stuffed animals. I told her that this would not be subtle or vague—Yogi was going to send a very obvious sign.

Yogi also showed me a significant piece of artwork. I asked if there was some sort of art or creation that had been made in Yogi's liking, and Julie said no, so I told her this may be something that showed up later. Yogi was also showing me how he was helping her and her boyfriend in all areas of their lives, even acting as a good luck charm for them. Yogi was cheering and celebrating from the other side and kept showing me a blue ribbon, like what someone receives as a prize in a competition. Julie told me she had just been awarded employee-of-the-year at her company and had been gifted Tiffany earrings in the well-known blue box with a big ribbon tied around it. Her boyfriend also had just won $6,000 in a poker competition. When Julie shared this with me, everything in my mind's eye got bright, almost like a strobe light going off, which is always a sign of confirmation. Yogi said, "Yes! See! I'm here and I am helping! I am so happy for you!"

Later that day, Julie e-mailed to tell me she shared the details of our session with a friend of hers. Her friend said she had commissioned a painting of Yogi and was going to surprise Julie with it. It had just been delivered and her friend wanted to bring it over. Julie was in complete shock and so was her friend, who laughed and said Yogi ruined her surprise.

A few days later, Julie was in the grocery store and saw a worker struggling with a box and a cart down an aisle. Something told Julie to go see what was happening, and the worker was dumping a box of stuffed penguin toys into the cart! Julie started to see penguins in random locations all the time, and she wasn't the only one. Her boyfriend, who worked in a restaurant with a casino, was suddenly very aware of a slot machine with penguins on it, though he had walked by that same machine many times without noticing it. He felt guided to sit down and played one round, instantly winning 100 dollars. Around the same time, Julie was thinking of Yogi and feeling sad the morning of Valentine's Day. She felt compelled to turn on the television, where her set was tuned to a feature on African penguins. The male penguins help gather materials for nest-building during this time of the year, and researchers provided them with cute paper Valentines that they then brought to their female counterparts as a bit of festive fun for preparing their nests. She knew at that moment that Yogi was sending her some love. These are just a few of the many synchronicities that Yogi guided his people toward from the afterlife.

In my experience, animals in the afterlife love to communicate with us and let us know they are with us. They love when we ask them for signs, and when we notice, acknowledge, and thank them for the signs, they send more! This is a great way to establish this new facet of your relationship with a loved animal, and over time it will strengthen your connection with them so they can help you in other ways.

Intuitive Communication Assistance

Animals in the afterlife are available to guide us not only in our own lives, but also in our communication with living animals. When I am in a session communicating with a living animal, I will often feel one of my own or one of the client's previous pets present in the session. They are there to assist the animal in opening up or me in understanding the messages. Sometimes they show up without me even asking, while other times I specifically call upon them for help.

William reached out to me after rescuing a five-year-old boxer named Ruffalo because he wanted help acclimating the pup into his new family and home. I could feel a male German shepherd wearing a Superman cape flying around William and Ruffalo. I asked William if he previously had a dog that fit this description, and he said yes, the dog's name was Clark Kent. Clark Kent showed me there were some issues with Ruffalo's leash, and William told me that Ruffalo had chewed through a leash and slipped out of his collar twice while they were walking. He had just bought a harness for Ruffalo—one with the Superman logo on it—but he hadn't used it yet because Ruffalo didn't seem to like having it on. I could feel that Clark Kent had guided William to that harness with his namesake's logo to help keep Ruffalo safe. Clark Kent also wanted William to call upon him for help in getting Ruffalo used to the harness, which, a month later, was no big deal for the energetic dog.

If they don't show up on their own as Clark Kent did, you can simply call upon the assistance of animals in the afterlife whenever you are intuitively communicating with living animals. Using any of the meditation practices shared in this book, simply add a step of setting an intention and asking for their assistance. For example, with the Great Glass Elevator technique (see page 120), after you ride the elevator and meet the animal you are communicating with, invite an animal in the afterlife to enter the space and assist as needed.

If for any reason you are not feeling them coming through, so be it; simply let it go. Sometimes there is no need for them to join. They have a higher perspective of the situation and may feel it is in the highest good for you and the living animal if they don't participate—perhaps so you have the opportunity to gain confidence without any assistance. If no one shows up, don't get distracted by it or take it personally. Their perspective may be, "You've got this! It's all you!" And they may join you to help on another occasion, so be open to asking whenever you feel guided.

Practice: Connecting with Animals in the Afterlife

Though you can certainly use any of the techniques shared in the previous chapter, here is another practice that is catered more specifically to connecting with animals who have passed into the afterlife.

1. Find a quiet and peaceful place where you won't be disturbed or distracted. If you wish, light some candles or incense to create a calming and relaxing atmosphere. It may help to hold a photo or personal item, such as a collar or toy that belonged to the animal. Close your eyes, take a few deep breaths, and focus your mind on the intention of connecting with them.
2. Imagine yourself stepping onto a soft, fluffy cloud. Visualize the cloud beneath your feet and feel the gentle support and buoyancy it provides. Look up into the beautiful blue sky and see other clouds floating above you. These clouds represent the spirit world.
3. Feel your cloud rising toward the spirit clouds, feeling the wind gently rushing past you and the sun shining down on your face. You may also hear the sounds of birds singing and feel a sense of peace and joy.
4. As you approach the spirit clouds, you see the animal waiting for you on one of them. They may appear in their physical form or as a radiant energy being. Your cloud meets theirs and you easily join them. Greet the animal with love and respect and let them know that you are here to communicate with them.

5. Use your intuition to listen to their messages and respond with empathy and understanding. You may receive images, feelings, or words that convey their thoughts and emotions. Take your time to talk with the animal and express any feelings or questions you have. Remember to listen attentively and be open to their responses. If you'd like, write down anything in your journal you feel guided to note.
6. When you feel ready to end the communication, thank the animal for their presence and connection. Visualize yourself stepping back on your cloud and slowly descending back to your space.
7. Take a few deep breaths and ground yourself by feeling your feet on the ground and your body in the present moment. You may also want to journal or meditate on your experience to gain further insights and clarity.

Spirit Animals and Power Animals

In many Indigenous cultures, spirit animals and power animals are seen as guides or helpers in the spiritual realm. Spirit and power animals are not specific animals that you have known in life; rather, they are archetypes and higher beings that appear as animals to you. These animals are believed to possess certain qualities or traits that can be used to provide guidance and support to humans.

A *spirit animal* is often seen as a representation of an individual's inner spirit or personality. It is believed that each person has a unique spirit animal that is with them throughout their life, and that this animal can provide insight and guidance on their spiritual journey. For example, on some occasions during animal communication sessions, I will feel a koala spirit animal present with me, reflecting and reminding me of my calm and nurturing energy and how I can help the animal I am communicating with. Sometimes I see the koala in my mind's eye, and other times I have an inner knowing and feel a sense of their presence and guidance.

A *power animal*, on the other hand, is a type of spirit animal that is called upon for specific purposes, such as healing, protection, or guidance in a particular area of life. It is believed that these animals possess a special energy that can be harnessed. As an example of a power animal, there have been times when I am intuitively communicating with an animal and feel the presence of a bear reminding me to listen to my instincts. I feel them bringing an essence of courage to help the animal I am communicating with feel more comfortable opening up to me.

Both spirit and power animals are often viewed as sacred and are treated with great respect in Indigenous cultures the world over. They are believed to be messengers from the spirit world and are seen as sources of wisdom, strength, and inspiration. Many Indigenous peoples, such as the Lakota Sioux tribe, believe that by connecting with their spirit or power animal, they can gain a deeper understanding of themselves and the world around them.

You may also call upon spirit and power animals to help you in your intuitive communication with living animals, like the bear sometimes does for me when I am working. Whereas animals that have crossed over feel lighter and almost more angelic, spirit and power animals give a larger, expansive impression and feel more advanced and complex, focusing on specific guidance or tasks. You do not have to choose one helper or another; you can invite any or all of them whenever you would like assistance.

Practice: Connecting with Spirit or Power Animal Guides

Before asking spirit or power animals for help communicating with living animals, it is important to meet and begin building a relationship with them. Here is a meditation process to help connect with your personal spirit or power guides.

1. Find a quiet and comfortable place where you won't be disturbed and close your eyes. Take several deep breaths, inhaling through your nose and exhaling through your mouth.

As you breathe, allow yourself to relax and release any tension in your body.

2. Visualize yourself in a peaceful natural setting, such as a forest, meadow, or beach. Imagine yourself surrounded by a warm, golden light that radiates from the center of your body and being. Feel the warmth of the sun on your skin, the breeze in your hair, and the earth beneath your feet.
3. Set the intention to connect with your spirit or power animal guide. Say a prayer or affirmation to invite your animal guide to reveal itself to you. You might say something like, "I open myself to the guidance of my spirit animal. Please reveal yourself to me and show me the way forward."
4. Pay attention to any sensations, emotions, or images that arise. Allow yourself to be open to any messages or insights. You might see an animal in your mind's eye, hear its unique call, or simply feel its presence.
5. Once you have connected with your animal guide, take some time to communicate with it. You can do this by asking it questions or simply sitting in its presence and listening to its wisdom. You might ask it questions like, "What message do you have for me?" or "How can I work with you more closely?" Listen to your intuition and trust the guidance that comes through. Take note of any insights or messages you receive in your journal.
6. When you are ready to end the meditation, thank your animal guide for its support. You might say something like, "Thank you for revealing yourself to me and sharing your wisdom. I am grateful for your guidance." Visualize the warm, golden light surrounding you once again, and slowly bring your awareness back to your physical body.
7. Take a few more deep breaths and open your eyes. Take some time to journal about your experience and any insights or messages you received. Write down any actions you want to take or changes you want to make based on the guidance you received.

The hardest part about having pets is that their lives are not as long as ours. But in my experience, our pets don't ever *truly* leave us. They are always with us in spirit, and they continue to be a loving and important part of our lives after they pass away, even guiding us in finding new pets to love and sending encouraging and supportive messages our way.

Next, we'll move on to more topics you can explore now that you've built a strong foundation in intuitive communication, including connecting with wild animals, communicating with more than one animal at a time, and information on auras and chakras.

Journal Reflections

1. Have you ever felt a connection with an animal in the afterlife? This might be through dreams, signs, or even physical presences, such as sensing they are near or seeing them out of the corner of your eye. What have these experiences been like?
2. What animals in the afterlife would you like to communicate with? What would you like to share with them? What would you like them to share with you?
3. How do you think the experience of communicating with deceased animals will feel compared to animals that are still living?
4. Spend some time exploring the idea of animal guides such as spirit or power animals. What kinds of animals do you already feel a strong connection with? What kinds of animals seem to pop up with synchronicity in your daily life? Write out some adjectives that describe the animal or symbolism associated with it and journal about how they might support your aims.

BEYOND THE BASICS

At this point, you have everything you need to communicate intuitively with animals. But as I have learned—and continue to learn—there is so much more to discover when it comes to our animals! This chapter discusses a variety of topics that you can explore once you have a strong foundation and have been practicing for a while, such as how to combine techniques or communicate with wild animals or more than one animal at a time. I also offer an introduction to some other topics that might be of interest, including auras and chakras and how they can be utilized in intuitive communication. Most of the information here only offers a starting point, lighting up just the first few steps of a bunch of potential paths. It's up to you to decide where your interest lies and what to pursue next.

Combining Techniques

While it may be tempting to dive right into multiple techniques, I recommend focusing your efforts on mastering one method first. As you know by now, intuitive communication takes time and doesn't always yield results. Trying multiple techniques can make it difficult for you to discern which method is actually working. Overall, I have found that a single technique approach increases effectiveness and leads to a more solid foundation, making it easier to integrate additional techniques later on.

One reason for this is confidence. This is a very personal endeavor, and it requires that you believe in and trust yourself. By concentrating on mastering one technique rather than spreading your attention thin, you learn more quickly, gain confidence in your abilities, and then boost those abilities, leading to more confidence—and so on. Once you feel confident in your foundation, you may then begin to explore combining techniques and methods to complement your existing skills.

For example, sometimes I will begin with a Body Scan (page 43) of the animal and then move into the Screen and Flower (page 117) technique. Another example might be taking the Great Glass Elevator (page 120) up to your Sacred Temple (page 122). Sometimes, while in the Sacred Temple, I then utilize the Heart-to-Heart (page 114) or Third Eye (page 115) techniques. There really is no right or wrong way of combining techniques. With practice, you will develop your unique way of communicating. But, especially in the beginning, I encourage you to keep it simple! Don't overthink or take on too much. Getting caught up in the process will distract you from being present in the moment for the communication itself.

Communicating with Multiple Animals

People often come to me wanting to know what their animals feel about changes in the household, which especially happens when new pets are introduced or when pets are not getting along. Recently, Eva contacted me because she was moving herself and her two pit bulls in with her mom. Her mom was nervous that her two cats would not get along with Eva's dogs. I communicated remotely with all four animals in one session to learn how each felt and to allow them to connect energetically with each other while I acted as a moderator, or pet therapist, if you will. I began by checking in with each animal individually to understand their experiences and perspectives. I could feel that the dogs had never met a cat before, but they were big softies and excited by the opportunity to have new siblings. I could also sense that one of the cats was very cautious and protective of her space while the other was curious and confident. She was encouraging the cautious cat to warm up to meeting the dogs. As we progressed, the cautious cat shared that as long as her space was respected, she would be willing to accept her new siblings. The dogs both felt very understanding and willing to abide by the queen cat's rules. When Eva and the dogs moved in, all went well, and they now live happily together.

As you become more skilled, it can be really fun to communicate with multiple animals at the same time.

Here are some tips for working on this more advanced skill:

- Before attempting communication with more than one animal, first establish strong foundations with each animal individually.
- Have clear intentions for how you would like everyone to engage. This will help you maintain focus and effectively tune in to each animal's energy. It is also important to create a safe energetic space for each of them through intention and visualization. When you feel calm, confident, and secure, the animals will too. This will encourage them to open up and share.
- Depending on the animals and the situation, you may feel inclined to visualize a protective energetic barrier around each animal. This will help to contain their energies and prevent them from blending or overwhelming one another, especially among animals that have not been getting along. That being said, I have never been in an animal communication session with multiple animals where they fought in any way that would be potentially harmful. The higher vibration of intuitive communication naturally provides a safe and supportive space.
- Visualization is a powerful tool to help you manage and organize the information you receive from multiple animals. You might try visualizing each animal's communication as distinct colors, shapes, or symbols to help you differentiate between the various energies.

- Maintain a balance between focus and flexibility. Stay present and attentive to the needs of each animal, while also being open to adapting your approach as necessary. This may involve shifting your attention between animals, refocusing your intentions, or employing different communication techniques with each.
- Active listening and validation are key. Make sure to give each animal your full attention and show them that their thoughts are valued. This can help to build trust and foster a stronger connection. You can facilitate a dialogue by asking open-ended questions and allowing each animal to express themselves. Be patient and allow the conversation to unfold naturally, ensuring that every animal has a chance to participate.
- If you encounter conflicting energies or emotions, remain calm and focused. Stay grounded, maintain your intentions, and approach these conflicts with empathy and understanding, aiming to mediate and find a resolution that benefits all parties involved. The animals will feel you as the leader in the situation, so by maintaining that feeling and frequency, you will have a successful communication session.

Sacred Round Table Technique

All of the communication techniques in Chapter 7 can be used when communicating with multiple animals. But I love this technique because it offers each animal their own space and is easier to facilitate. It almost reminds me of a conference room meeting, but with a loving, supportive twist.

1. Begin by grabbing your journal and finding a quiet and peaceful location (such as your sacred space) where you can focus your attention without any distractions.

2. Set your intention before starting by visualizing a positive, loving, and harmonious connection with the animals. Do a grounding and centering practice of your choice to clear your mind, focusing on your breath and letting go of any stress or tension.
3. In your mind's eye, visualize a sacred round table in front of you. This table should be large enough to accommodate all the animals you wish to communicate with. Imagine the table is surrounded by a soft, warm light, creating a welcoming space.
4. Invite the animals to join you at the table one by one. As you do this, visualize each animal taking their seat and becoming a part of the circle. Acknowledge their presence and thank them for joining. Focus on establishing a connection with each of the animals. Allow yourself to be open to any thoughts, feelings, or images that may come to you.
5. Once you feel connected, initiate a dialogue. Ask questions, express your thoughts, or simply allow the animals to share their perspectives. Be patient and allow each to communicate, ensuring that every voice is heard. Write down any information that comes to you.
6. Gently bring the session to a close by expressing your gratitude for their participation. Visualize the sacred round table dissolving into the surrounding light.
7. Slowly come back to your physical surroundings, taking a few deep breaths to ground yourself back in your space. Spend some time reflecting on the communication session and review what you wrote down in your journal for insights.

The Journey of Animal Souls

Throughout each lifetime, the soul both learns and teaches lessons before crossing over to the afterlife—sort of like how humans advance through school. We learn and grow through one grade and then graduate to the next level. Each lifetime for the animal is like a grade; when it's complete, they move to the afterlife and then come back to an earthly experience in a new body. But the soul has advanced and grown from the lessons and experiences in the previous life.

When tapping into the energy of animals, I often feel the essence of their souls. Some are younger, some older, but all exhibit distinct qualities and characteristics. Younger animal souls are enthusiastic and energetic, and they tend to feel curious and eager to learn from their surroundings. Play is a natural way for them to learn and grow. Younger souls may engage in amusing and entertaining activities to satiate their curiosity and develop their skills. They can be emotionally sensitive and may require more attention, reassurance, and nurturing from their caregivers. Their emotional vulnerability can also make them more susceptible to stress, anxiety, or fear. As they have not yet had many incarnations, younger souls may come across as inexperienced and may be more prone to making mistakes or struggling with problem-solving. Despite this, they are willing to learn and grow from their experiences. They are usually eager to please and will actively seek guidance and validation from their caregivers.

Older animal souls feel wiser and more evolved because they have experienced multiple incarnations and accumulated knowledge from their past lives. They have a deep sense of calmness and serenity and can come across as more emotionally stable, displaying an innate wisdom that allows them to remain composed in various situations. Having experienced many incarnations, older souls develop a deep understanding of the emotions and needs of others. They are empathetic and compassionate, often showing great care and concern for the well-being of their caregivers and other animals. They are also more in tune with their intuition, the energy of others, and the spiritual realm. They understand the impermanence of life and can navigate through different situations with grace and resilience. Older souls often serve as teachers or guides, helping younger souls to learn and grow, whether it be here in our earthly lives or when they are between lives in the afterlife. They often provide valuable life lessons and spiritual insights to both animals and humans in their lives.

All that said, all animals are not necessarily either a younger or older soul. Many animals I communicate with feel somewhere in the middle, with aspects of both ends of the spectrum depending on their experiences. But getting a sense of an animal's soul age gives me insight to understand them better.

For example, Lisa has two chihuahuas from the same litter named Tia and Sola. When I met them, they were three years old. I could feel that Tia was an older soul and Sola was a younger one. Lisa took them to a doggie daycare where Sola was always running around, barking, and stirring things up with the other dogs. She was a feisty little one, barreling into everything and everyone. She was loving but very vocal, always jumping into situations without thinking, and she had a hard time learning from her mistakes, such as repeatedly digging in the garbage and going to the bathroom in the house.

Tia, on the other hand, would always find an elevated dog bed out of reach of the others at daycare. She preferred not being a part of the riffraff but had no judgment about those who participated. She had a grounding presence that would often calm other dogs when they were close to her. Tia always knew when Lisa was sad or not feeling well and would lie with her, knowing her energy offered support. Tia understood Sola and loved and accepted her even though she was always causing trouble. When I shared these perspectives with Lisa, it deepened her respect for Tia and helped her to accept Sola for who she was rather than comparing her to Tia.

By recognizing and honoring the unique spiritual qualities of young and older animal souls, we can facilitate deeper connections, which leads to a greater appreciation of the spiritual aspects of the human-animal bond.

Lost Animals

When an animal goes missing, it is extremely stressful. While intuitive communication can help provide insights into the animal's emotional state, physical well-being, and location, it can be challenging for several reasons. First, both the lost animal and their guardian may be experiencing intense emotions, making it difficult to establish a clear and focused connection. It is essential to remain calm and centered during the process, allowing the emotions to flow through you without becoming overwhelmed. In addition, the animal may be in an unfamiliar environment that is taking up their focus, making it harder for them to receive and respond to intuitive communication. Finally, external factors, such as thoughts and energy from the animal's caregiver or others, can also interfere with the energetic connection. To minimize interference, practice grounding techniques and set a strong intention before initiating contact.

The goals in establishing an intuitive connection with a lost animal are to check on their well-being and receive information to locate them. Once grounded and ready to begin, I will pull up the location of where the animal was last seen on a digital map. Then, I close my eyes and tune in to the animal, asking questions about what they see, smell, hear, and feel. I will then open my eyes and look at the map to see what locations are "lighting up" for me—that is, where my attention is drawn. Then I can tune in deeper to those areas and ask more questions.

In my experience, lost animals often share information about where they were recently rather than where they currently are. Because their adrenaline is elevated from being out of their element, their communication can be all over the place. Asking specific questions about surroundings, landmarks, and any people or animals they may have encountered can help narrow down their location and guide search efforts.

One lost dog case I helped with was for Humphrey, a beagle mix who had been spooked by fireworks and ran away from his backyard. I pulled up Humphrey's last known location as I was tuning in to him. I immediately felt how highly elevated his energy was; he was in total flight mode. When I asked him to share his surroundings, I saw a small body of water, but I couldn't find one on the map. I also saw a small brown brick building, what looked like a square with lit up digital information on it, and something like a yellow triangular flag that possibly had bumblebees on it. I communicated to Humphrey to come out of hiding and reveal himself to anyone nearby. The area northeast of his house was lighting up on the map for me, so I encouraged Julian, Humphrey's guardian, to post flyers in that area.

The very next day, Julian got a call from a teacher at a nearby school—a brown brick building with a digital sign out front. She said that Humphrey came out of the bushes from behind the school and laid down in front of her, exhausted. She had seen Julian's flyers and called him immediately. When Julian arrived at the school to pick up Humphrey, he saw yield signs that alerted drivers to the crossing zone in front of the school. They didn't have bumblebees on them, but they did have yellow and black stripes, just like a bee! This is an example of why it's important to be flexible in your interpretations.

Locating a lost animal can be a challenging, emotionally draining, and time-consuming process. I always make sure to manage expectations so that everyone I am helping is aware there are no guarantees. If the initial communication session doesn't lead to the lost animal's location, you might need to repeat the process, refining your focus, and asking for more specific details. Regardless, be sure to pursue all other avenues for finding a lost pet: putting up flyers, informing neighbors, calling local shelters. If possible, I recommend collaborating with trained search-and-rescue teams as well. There have been times when animals I have worked with were found quickly, other times when the animal was found several weeks or months later, and sadly, times when they were never found. For lost animal cases, I always come prepared to just do my best and surrender to whatever the outcome may be.

Wild Animals

Communicating with wild animals can be a thrilling and humbling experience. Overall, while there may be some differences in how intuitive communication with wild and domesticated animals occur, the fundamental principles remain the same. With that in mind, I will walk you through some of the unique challenges wild animals present.

First, the foundation of any successful communication with wild animals lies in respect. Wild animals are generally guarded and wary of humans—with good reason—so it is essential to energetically approach them with humility. A key to establishing trust and respect is understanding the animal's natural instincts, behaviors, and body language. For example, some wild animals may rely more on body language to communicate, while others may be more vocal or use scent marking. Domesticated animals that we have formed emotional connections with, such as cats, dogs, and horses, are more familiar with human presence. Often, they are trained to respond to specific cues, which means they have a certain understanding of communication with humans. With wild animals, we don't have that foundation, and building trust can be a more challenging and time-consuming process.

When attempting to communicate with wild animals, it's also crucial to consider their environment. Animals in their natural habitat are more in tune with the subtle energies and rhythms of nature. To connect with these animals on a deeper level, we need to attune to the energy of their environment too. In my experience, wild animals have a more rapid energetic vibration, due to how they are always on guard and aware of their environment. By being present and mindful of your thoughts and emotions, you can create an inviting, calming energy for them.

When I communicate with wild animals, whether it be with squirrels, raccoons, or even bears, mountain lions, or wolves, I connect with them using the Vibrational Match technique (see page 119). This works well because it focuses on finding and connecting with the animal on their level, where they feel safe. Not long ago, there was a cheeky squirrel who would frequently visit the garden in my backyard. I noticed, however, that the squirrel often ran a bit too close to where my three dogs played, and I was a bit concerned for its well-being and did not want it to feel traumatized or in harm's way.

One sunny afternoon, I saw the squirrel perched on a tree branch near our patio. Seizing the opportunity, I sat quietly in the garden and focused on becoming a vibrational match with the squirrel. I sent gentle, loving energy toward her and envisioned a light emanating from my heart, reaching out to the squirrel, symbolizing my peaceful intentions. I proceeded to communicate a message of safety and caution to the squirrel. I also sent images of my dogs and the areas they frequented, followed by feelings of alternative routes for the squirrel to take through the yard. In addition, I also sent feelings of welcome and safety associated with the other areas of the garden, out of the dogs' range. As I finished communicating my message, the squirrel was looking directly at me, its tiny eyes full of understanding. I vibrationally felt her acceptance of and agreement with my message. Then she chittered softly, flicked her tail, and scampered away.

Over the next few days, I noticed a change. The squirrel began to steer clear of the areas where my dogs played, while still visiting other areas of the garden. One day, to my surprise and delight, the squirrel appeared on my garden wall, then jumped on the ground and came right up to me within a few feet—and she wasn't alone. There, peeking from behind its mother, was a baby squirrel, its eyes wide and curious. The adult squirrel seemed to display her young one proudly, and at that moment, I felt a wave of gratitude and connection wash over me.

This encounter was a powerful reminder of the deep bonds that can be formed through intuitive animal communication. It reinforced the notion that animals understand more than we give them credit for, and with respect, love, and a bit of patience, co-existing in harmony is entirely possible.

Auras

Auras are ethereal energy fields that surround all living beings and serve as a reflection of an individual's emotional, physical, and spiritual well-being. The concept of auras is found in numerous cultures, with mentions in various religious and spiritual texts from Hinduism to ancient Egyptian belief systems. While the idea of an energy field surrounding living beings is not a new one, the scientific understanding of these energy fields has evolved.

In the 1930s, Russian inventor Semyon Kirlian discovered a way to capture images of auras using high-voltage photography, now known as Kirlian photography. This technology led to further research, and modern science now suggests that these energy fields result from the electromagnetic energy emitted by living beings, including people, animals, and plants.

Reading an animal's aura can provide valuable insight into the animal's emotional and physical state, including stress or fear they are experiencing or areas of their body that may be in pain. It can also offer details about their personality traits, including what might be triggering behavioral issues, and past experiences, especially those that were traumatic.

It's important to note that the aura is always changing. Just like our emotions and physical state fluctuates throughout the day, so does our aura. While aura photography can be interesting, I find that better insights come from interpreting the aura in real time rather than relying on static readings.

Animal auras, like human auras, can exhibit a wide range of colors, each with a unique meaning. As a general reference, here are the most common aura colors and their interpretations. If you find that different qualities line up with these colors—that's great! There is no right or wrong here. Feel free to take notes on your associations and create your own record of aura characteristics.

- *Red* represents strong energy and vitality, but it may also be a sign of aggression, reactivity, or anxiety.
- *Orange* indicates creativity, enthusiasm, and sociability. It can also signify a strong emotional connection.
- *Yellow* denotes intellect, curiosity, and a playful nature. It may also indicate a strong bond between the animal and their caregiver.
- *Green* is indicative of healing energy and a strong connection with nature. Animals with a green aura are often very nurturing and compassionate.

- *Blue* signifies communication, loyalty, and a calm temperament. These pets are often excellent listeners and can sense their human's emotions.
- *Indigo* represents intuition, sensitivity, and psychic abilities. Pets with an indigo aura are often deeply connected to their humans and can sense their energetic and emotional state.
- *Violet* is associated with spirituality, wisdom, and divine connection. Animals with a violet aura are often considered old souls and may have a calming effect on others.
- *Gray or black* can be associated with blocked energy, such as the animal being in fog and state of confusion, or potential health issues.

Practice: Reading Animal Auras

1. Find a quiet, comfortable space free from distractions. Close your eyes and take a few deep breaths. Visualize a grounding cord gently attached to the base of your spine and extending down to the core of the earth, grounding your energy.
2. Whether you are in person or using a photo, take a few moments to observe the animal's behavior, body language, and overall demeanor. Release any preconceived notions or judgments.
3. Establish a connection by gently sending out your energy and intentions. Visualize a beam of light or energy extending from your heart to the animal's heart. Or you may feel guided to use one of the other techniques (see Chapter 7). Approach the moment with an energy of respect and honor for the current frequency of their aura.
4. Close your eyes and visualize the animal surrounded by a glowing energy field. Notice any colors, patterns, or intensity variations. Allow yourself to feel whatever comes

up as you tune in to the energy field. Write down whatever comes through as you feel guided to do so.

5. If you are in person, you may also use your hands to gently feel the energy surrounding the animal without physically touching them. If using a photo of the animal, you can run your hands over the photo with the intention of feeling their aura. Pay attention to any sensations, such as warmth, coolness, or tingling, which may indicate the presence and state of their aura.
6. When you've finished your reading, thank the animal for sharing their energy with you. Gently disconnect your energy from the animal and ground yourself once more to ensure a clean disconnection.
7. Spend some time referring back to your notes and interpreting what you saw and felt from their aura to better understand the animal's needs, emotions, and overall well-being. Often more insights arise when reviewing the notes after a session.

Chakras

The concept of chakras may be familiar to those who practice yoga or meditation or engage in other holistic healing practices. *Chakras* are energy centers within the body that correspond to specific organs, glands, and areas of the physical and emotional body. They are not physical locations but rather metaphysical entities that exist on an energetic level. In Sanskrit, the word *chakra* means "wheel," signifying the constant spinning motion of these energy centers.

Just like humans, animals have a chakra system. Understanding and working with an animal's chakras can open up an entirely new level of intuitive communication, helping to deepen bonds. This topic could be a whole book in itself, but here are a few tips regarding chakras and animal communication.

There are seven primary chakras:

- *Root Chakra:* Located at the base of the spine, it is associated with security, survival, grounding, family, the pack, and ancestry.
- *Sacral Chakra:* Positioned just below the navel, it governs creativity, emotions, and sexuality.
- *Solar Plexus Chakra:* Located above the navel, it is linked to personal power, self-esteem, and decision-making.
- *Heart Chakra:* Positioned at the center of the chest, it is responsible for love, compassion, and empathy.
- *Throat Chakra:* Located in the throat, it is connected to communication, self-expression, and truth.
- *Third Eye Chakra:* Positioned between the eyebrows, it governs intuition, psychic abilities, and inner wisdom.
- *Crown Chakra:* Located at the top of the head, it is linked to spiritual connection, enlightenment, and universal consciousness.

Animals and humans share these seven primary chakras, but the energy flow and distribution may differ depending on the species. For example, the root chakra in animals is generally more developed than in humans, as they rely heavily on their instincts for survival.

I will often scan an animal's chakras to see if any feel out of balance. If, in my mind's eye, the color of the chakra appears muted or dimmed or if it is spinning slowly, that tells me that area is challenged. For example, the throat chakra is associated with communication and self-expression. If I am feeling a block or imbalance there, it could mean the animal is not getting the opportunity to express themselves, which is adversely affecting them. Or there could be a health issue in that area of the body, such as a sore throat or other complications.

Practice: Chakra Scanning and Balancing Meditation

Here's a simple guided meditation you can use to scan the chakras of an animal to gain insight about their well-being, and even work to balance them.

1. Find a quiet and comfortable space where you and the animal can relax without any disturbances. If you are connecting with the animal remotely, have their photo in front of you. Close your eyes and take a few deep, slow breaths to center yourself.
2. Begin by visualizing a warm, healing light surrounding both you and the animal. This light will serve as a conduit for supportive energy. Now you will go through each of the chakras, noticing any changes in their color or speed of rotation as well as what you are intuitively feeling for each. Take notes on anything that stands out to you. If you wish to balance any of the chakras, pause on it and imagine a vibrant version of that chakra's color. Breathe in and visualize that energy becoming brighter and spinning freely at the perfect speed. As you breathe out, imagine any tension or blockages fading away.

    - Focus your attention on the Root Chakra, a vibrant red energy located at the base of their spine.
    - Shift your focus to the Sacral Chakra, a warm orange energy located in the lower abdomen area.
    - Move your attention to the Solar Plexus Chakra, a radiant yellow energy located in the middle of the abdomen.
    - Focus on the Heart Chakra, the beautiful green energy located in the center of their chest.
    - Shift to the Throat Chakra, the soothing blue energy located at the throat.
    - Move your focus to the Third Eye Chakra, a wise deep indigo energy located between their eyes.

- Finally, home in on the Crown Chakra, a lustrous violet energy swirling at the top of the head.

3. Take a few more deep breaths, visualizing the harmonized and balanced chakras as a whole within the animal, glowing with their respective colors. As you exhale, imagine any residual energy being cleansed and released from both you and the animal.
4. When you feel ready, gently open your eyes and bring your awareness back to your space. Review any notes you took to gain more insight into what you observed and felt.

Journal Reflections

1. What has been your experience combining different communication techniques? Did you find it more beneficial to stick with one technique or combine them? How so? Describe your experiences.
2. What kinds of wild animals live in your area that you could begin cultivating relationships with? What would you like to learn from them? How is your experience communicating with wild animals similar or different from domesticated ones?
3. What did you learn about an animal after tuning in to their aura? What colors did you see and feel? What thoughts, feelings, and emotions were associated with the color?
4. What did you learn about an animal when scanning their chakras? How did accessing the chakras help your communication and understanding of them?

Part III

PUTTING IT ALL INTO PRACTICE

BUILDING UPON YOUR FOUNDATION

Now that we have extensively covered strengthening our intuitive muscles, energetically and telepathically connecting with animals, and the various ways of communication, we move forward intending to make animal communication a lifestyle—more second nature and instinctual, just as communicating with other humans is—as well as to take it out into the world.

Intuitive animal communication requires patience, dedication, and practice. It's a journey that allows you to build a deep connection with animals, understand their emotions, and even be an advocate for their needs. Staying committed to this practice is essential for refining your skills. In this chapter, we will explore various strategies to help you cultivate your intuitive life with animals.

Staying Committed

When we first try something new, it is exciting and often revitalizing, bringing us fresh ideas and waves of good energy. But as we settle into a routine with anything—whether it's a new job or a new pair of shoes—the shine wears off and it becomes another expected part of our experience. Losing some steam around something you're learning is natural, but when it comes to animal communication, it will help if you find ways to remain committed. Here are some tactics you can try to make intuitive practice part of your daily routine.

Engage with Your Journal

I find journaling to be a valuable tool in many aspects of life, and especially so in animal communication. Journaling is an interpretive art; the meaning of what you write while connecting with an animal is not always immediately obvious, and signs may take a while to materialize. Of course, consciously journaling also allows you to track your progress, identify patterns, and reflect on your experiences. As I have noted throughout the practices, make sure to note specific details (such as the animal's body language if communicating in person), your intuitive impressions, and any messages or images you receive during sessions. Afterward, write down your thoughts, feelings, and intuitive insights. Don't judge or second-guess yourself. Just write down whatever comes.

I also recommend setting aside time to review your journal entries periodically. Regular reviewing can provide valuable insights about how your skills are developing and can serve as a reference for future sessions that fuel your growth. Reflect on your progress and identify patterns or trends to inform your future communication efforts. Reminding yourself of how far you've come is one of the best kinds of motivation.

As you review your journal entries, ask yourself questions like:

- What challenges have I faced in my animal communication practice, and how have I overcome them?
- What successes have I experienced, and what factors contributed to those successes?
- Have I noticed any patterns or recurring themes in my communication sessions?
- Are there any areas where I can improve or refine my skills?

Celebrate Your Wins

Celebrating your successes is a vital part of staying committed. Recognizing your achievements, no matter how small, can boost your confidence and encourage you to continue refining your skills. By acknowledging your successes, you cultivate a positive mindset and reinforce the positive actions that led to those achievements.

For example, if you successfully communicated with a nervous animal and helped them feel more at ease, take a moment to celebrate this accomplishment. Or maybe you communicated with someone else's animal, and you received information that the person was able to validate. Recognizing this success reinforces the techniques you used, helping you build confidence and improve your skills over time. When you celebrate your wins, you also ignite your enthusiasm and drive to keep practicing. Each success, no matter how small it seems to be, can serve as a source of motivation.

This will also help you stay resilient in the face of challenges and setbacks, keeping you committed to your practice. In moments of doubt or difficulty, remind yourself of your past successes and the progress you've made. You have everything you need to keep getting better.

Revisit Your Goals

As you progress through this journey, your intentions and goals may evolve. Regularly checking in with your intentions ensures that your practice remains aligned with your personal development. Revisit your original reasons for strengthening your intuition and assess whether these intentions still resonate with you or if they need to be adjusted.

Consider the following questions:

- What initially inspired me to learn animal communication?

- How have my experiences influenced my thoughts about animal communication and what is possible with it?
- Do my original intentions still align with my current values and aspirations?

If your original intentions no longer align with your current goals, that's perfectly fine. Just set new ones! This alignment will help you stay committed and focused on your practice. For example, you may have initially pursued animal communication to better understand your pets, but as you've grown in your practice, you may now feel called to help others connect with their animals as well. Adjust your intentions to reflect this new goal and use it as a guiding force.

Reflect on Your Energy

Energy, as you know, plays a crucial role in intuitive communication. Connecting with and gaining insight into your energy allows you to become a more effective communicator. Consider the following strategies for reflecting on your energy:

1. *Assess emotions and thoughts:* Before each animal communication session, take a moment to check in with yourself and evaluate your current emotional state. If you are feeling anxious, stressed, or distracted, take some time to ground yourself and cultivate a more balanced, focused energy before engaging with animals.
2. *Cultivate positive energy:* Practice mindfulness, meditation, and other techniques daily. By building up the positive and empathetic energy that these practices cultivate, you create a supportive environment for intuitive communication, allowing for deeper connections.

3. *Evaluate your methods:* Regularly assess your communication methods and their impact on your energy. Make adjustments as needed to ensure you maintain a positive, focused, and clear energy during your interactions with animals. Pay attention to the methods and techniques that resonate with you and produce the most positive results. If you find that certain methods consistently leave you feeling drained or disconnected, consider exploring alternative approaches.

Practice, Practice, Practice!

Committing to regular practice is essential for refining your intuitive animal communication skills. Especially in the beginning, I recommend blocking off time in your schedule or calendar to practice. This can feel like just another thing to add to your already busy life, but scheduling dedicated time demonstrates your commitment and holds you accountable to yourself and your practice.

Here's how to make it work:

1. *Find regular practice time:* Determine how much time you can commit to your animal communication practice each week. Consider factors such as work, family, and personal commitments. Be realistic about your time and set attainable goals to prevent burnout and maintain a healthy balance. It is also fine to take your practice schedule for a test drive; for example, you might commit to trying three 30-minute sessions a week for a month, and then reevaluate when the month is up.
2. *Create a practice routine:* Establish a routine for your practice sessions, including warm-up exercises, meditation, and journaling. A consistent routine will help you make the most of your practice time

and create a sense of structure and familiarity that supports your growth. For example, you might begin each session with a grounding meditation, followed by a series of visualization exercises to help you connect with your intuition before connecting with an animal. After your communication session, document your progress and reflect on any insights or challenges in your journal. If you have to be strict with your time, set a timer for each part of your session so you can settle in without feeling like you need to keep checking the clock.

3. *Find supportive community:* Having an accountability partner to check in and discuss your progress with can provide motivation and support, helping you stay committed to your practice. Make sure this person is open-minded and interested in your intuitive practice so they can be a positive voice. You may even consider connecting with fellow people exploring their intuition and animal communication by attending workshops or classes or seeking out mentors. Surrounding yourself with people and resources that support your growth and development will help you thrive.

Remember to have fun and enjoy the journey! Whenever you feel pressure around the commitment to your practice, take a breath and pivot toward the positive—what feels good and what you love about everything you are learning and gaining.

Letting Go of Perfection

When we are kids, we spend a lot of time exploring the world, asking questions, and learning through trial and error. As young children we aren't expected to have all the answers and perform perfectly all the time, but by the time we are considered adults, we have been instilled with this societal belief that we should be perfect. The truth is, we don't have all the answers, and we never will. That is the joy and wonder of the human experience, and if we open ourselves up to it, the world can be a place of constant learning, innovation, and variety. To find our way to this humbly awe-filled existence, we have to let go of the idea that perfection is achievable or even something that we should strive to attain in the first place.

Embracing the learning process is essential. True growth and improvement require patience and the willingness to learn from your experiences. Be open to receiving feedback, both from the animals you communicate with and other people, and use this feedback to guide your growth. And remember to be gentle with yourself: growth takes time, and it's normal to encounter challenges and setbacks along the way. Treat yourself with kindness and compassion, recognizing that learning and improvement are ongoing processes. If you show up and do the work, positive results will come. As a reminder, prioritizing self-care will help you be resilient.

Beginner's mind is a concept that comes from Zen Buddhism referring to an attitude of openness, eagerness, and a lack of preconceptions when studying a subject, just as a beginner would. The idea is to maintain a sense of curiosity and openness to new possibilities rather than becoming closed off or fixed in your beliefs as you gain expertise. It's the difference between thinking you know everything about a subject and being open to the idea that there's always more to learn. In the context of intuitive animal communication, beginner's mind can be especially valuable. Animals do not communicate in the same ways that humans do, and different species have different ways of expressing themselves. By approaching each interaction with an animal with a beginner's mind, you can be more receptive to the unique ways that individual animals might communicate.

Perhaps as a culture we are so obsessed with the fallacy of perfection because failure is seen as a bad thing to fear and avoid, rather than something that is just part of the process. If I have one overarching message for this book, it's that *practice* should be your main aim, in intuitive animal communication as in any other area of life. It's natural to feel nervous or fearful as you try new things; the important thing is not to let those emotions stop you from trying.

Here are some strategies to help you overcome nerves and release fear:

- *Acknowledge your emotions*, understanding that fear and nervousness are natural parts of the learning process and that it's normal to experience them.
- *Reframe your perspective* from nerves to curiosity and excitement. Embrace the opportunity to learn, grow, and connect with animals more deeply. This is when relaxation techniques are essential—practice deep breathing or other meditative techniques before and during your animal communication sessions.
- *Discuss your fears* with your support system. Whether you have one supportive accountability partner or are a community of fellow animal communicators, other people can provide encouragement, guidance, and reassurance.

Maintaining and Honoring Boundaries

In Chapter 2, we focused extensively on energy recognition and exploration. With practice, this becomes something you are innately aware of and always tapping into instinctually. Therefore, maintaining energetic boundaries becomes essential for several reasons:

1. *Protecting your energy:* Engaging in intuitive communication can be energetically demanding. By establishing energetic boundaries, you prevent your energy from being drained, ensuring that you can sustain your practice and still live your life without experiencing burnout.

2. *Preventing emotional overload:* This process involves tapping into the emotions, thoughts, and experiences of animals. Energetic boundaries allow you to empathize with animals without becoming overwhelmed by their emotions, which ensures that you can effectively support them without compromising your well-being.
3. *Supporting clear communication:* Energetic boundaries help you distinguish between your own emotions and thoughts and those of the animal. This clarity is essential for accurate and effective communication. By maintaining clear boundaries, you can better interpret the information you receive from animals and guide their guardians.

Another boundary that's key is respecting the privacy and autonomy of the animals you communicate with, as well as their caregivers. Seek consent before engaging in communication—animals, like humans, have a right to privacy. A teacher of mine once said, "Never offer unsolicited advice," and I feel these are words to live by. For example, sometimes I am out walking and feel communication coming from a dog who is walking with their human, a person I don't know. Even if I am feeling stress or some sort of negative energy, I am not going to walk up to the person and tell them what I am picking up. It is none of my business. It can be challenging to do nothing in situations where you feel so much from an animal and their caregiver. And certainly, there will be times when you are guided to take action in some way. But I understand I am not here to fix everyone. I do my best and show up where I am guided so I can serve the highest good for all involved. But feeling the pressure to save everyone can be extremely draining and will only hold me back from truly serving.

Here are some strategies for setting and respecting boundaries:

- *Ask for permission* from people and their animals before tuning in to an animal's energy. This respects their personal space and privacy, allowing for a more positive exchange. With animals, this can be done by simply stating your intention intuitively, for example, "May I communicate with you?"

- *Practice energetic shielding,* a technique where you visualize a protective layer around yourself that keeps out unwanted energies (see following practice).
- *Grounding* exercises help manage energy flow. Connecting to the energy of the earth can drain excess energy that might make you susceptible to unwanted communications.
- *Be mindful* of where you focus your attention. Just because you can tune in doesn't mean you should. It's important to remember that not every situation calls for intuitive communication.
- *Self-care* for your physical, emotional, and mental selves is crucial. When you are well-rested and balanced, you are less likely to be affected by external energies.
- *Set specific times* for your intuitive work with animals. This not only helps create a boundary but also signals to the animals when you are open to communication.
- *Consciously disconnect* after communicating, such as by visualizing a cord being gently disconnected or cut.

Practice: Energetic Shielding

Shielding is an energy work technique that protects you from external energies that you may encounter in your daily life. It's not meant to block out all energy exchanges, but rather to filter out the ones that might be unhelpful, overwhelming, or draining. Shielding is not about fear or blocking out all connections. It's a loving act of self-care that allows you to engage with the world, and particularly with animals, in a balanced and respectful way. It allows you to provide your compassionate services without taking on any unnecessary energetic burdens.

1. Before you begin engaging with the world, take several deep, calming breaths. This helps to center your energy and prepare for the shielding process.
2. Imagine a bubble or light around your body. This shield can be any color that resonates with protection and positivity for you. Some people envision a white light for its purity, others may choose a color like pink (as discussed in Chapter 4), or whatever other color relates to protection.
3. As you visualize this shield around you, set a clear intention that it is there to protect you from any negative or uninvited energy. You might say to yourself, "This shield protects me from any energy that is not for my highest good."
4. It may be helpful to add an extra step to empower your shield. You might imagine a symbol of power or protection (like a star, spiritual symbol, or anything that feels strong to you) overlaying the shield.
5. Keep in mind that this shield isn't permanent and might need to be refreshed periodically. You might choose to renew your shield every morning, or before engaging in any energy work.

Compassion Fatigue

This line of service comes with its own set of challenges, one of the most significant being compassion fatigue. *Compassion fatigue* is a condition characterized by emotional, physical, and mental exhaustion arising from continuous exposure to the emotional pain and suffering of others. It is often referred to as "the cost of caring" and can be accompanied by a diminished capacity for empathy and compassion for others. If we don't practice self-care and maintain energetic boundaries when doing this work, we may start feeling drained, exhausted, irritable, or even apathetic. Compassion fatigue can also impair a person's ability to effectively tap into their intuitive abilities, further exacerbating feelings of frustration.

Working with animals who have experienced traumatic or distressing situations, such as cases of abuse, neglect, or severe illness, can take a significant emotional toll, contributing to the development of compassion fatigue. If we tune in to an animal that is suffering in some way, we may start taking on their experience or feeling a sense of responsibility for their well-being—more reasons why energetic boundaries are so important. It is very important to be able to separate yourself from the suffering you encounter. Be aware that certain personal factors can increase an individual's vulnerability to compassion fatigue, including a history of trauma, unresolved personal issues, or a lack of attention to one's health and well-being.

Here are a couple of tips for managing compassion fatigue:

- *Regularly assess your own emotional state* to develop self-awareness around the signs of compassion fatigue so you can take proactive measures to manage your emotional well-being (see Chapter 4).
- *Create and maintain healthy personal boundaries*, including setting limits on the number of emotionally challenging situations or cases you assist with, scheduling regular breaks or vacations, and prioritizing personal interests and hobbies that feed your spirit.
- *Tap into your support network*—it bears repeating! It is always good to maintain a solid connection with other people who understand the challenges you experience and offer emotional support, validation, and guidance in navigating difficult situations.
- *Incorporate regular self-care and stress-reduction practices* including meditation, yoga, exercise, journaling, or engaging in creative pursuits that provide a sense of relaxation and fulfillment.

Practice: Dissolving Compassion Fatigue Meditation

By practicing this meditation, you can strengthen your emotional resilience, replenish your energy, and maintain a healthy balance between compassion for others and self-care.

1. Find a peaceful environment where you can sit or lie down comfortably without distractions.
2. Close your eyes and begin taking slow, deep breaths. Inhale deeply through your nose, filling your lungs completely, and then exhale slowly through your mouth. Concentrate on the sensation of your breath entering and leaving your body.
3. As you continue to breathe deeply, scan your body, starting from the top of your head and moving down to your toes. As you scan each area, imagine the energetic cords you have that are attached to all the situations in your life that may be emotionally draining you—animals, people, places, situations, traumas, and so on.
4. Visualize a bright, beautiful, healing light high up in the sky. It is shining down upon you and all around you. This light represents a source of energy that can replenish and restore you. As you inhale, imagine this healing light entering your body and filling every cell with revitalizing energy.
5. As you exhale, imagine any negative emotions, stress, or fatigue being released from your body, leaving you feeling lighter and more balanced. Continue to breathe in the healing light and release negativity with each exhale. With each breath, all the cords that connect you to what is draining begin to dissolve.
6. Silently repeat a self-compassion affirmation, such as, "I release, realign, and reset" or "I give myself permission to rest and replenish."
7. Continue to bask in the healing light until you feel the cords have dissipated, and your whole body is filled with beautiful, supportive light.

8. When you feel ready, gently bring your awareness back to your body and the physical space around you. Feel the connection between your body and the ground beneath you, allowing yourself to feel rooted, secure, and peaceful. Slowly open your eyes and take a few moments to reorient yourself to your surroundings.

Managing Expectations

People often come to me hoping to find a quick fix for behavioral issues they are experiencing with their pets. They want me to tell their animal how to behave, such as asking their cat to stop using the couch as a scratching post. I can intuitively communicate to them what behavior is appropriate and what is expected of them. And they may respond with, "Yes, I get it!" But every moment is a new moment, especially for an animal. They make choices based on the energy in that present moment, which may contrast with a message you previously shared with them.

The next day, that very same cat may get caught up in the moment and revert to the habit of scratching on the couch. I always encourage maintaining energy, communication, and connection so the animal deeply absorbs the message and starts to make appropriate choices. It can be a process to get the message to stick, so patience and consistency are key.

Also, just because the animal understands what we are telling them does not mean they are going to listen. In my experience, some animals get the point quickly, but usually it requires holding space and reminding them intuitively and energetically. This can be frustrating, but set an intention to notice that feeling, bring it out of the shadows, and release it (such as through Tapping, see page 50). You don't have to spend a lot of time focusing on the frustration, but even a few minutes of venting out your feelings and frustrations in your journal can help release the energy.

When considering expectations, there may be times when certain information is not revealed during a session. This can be attributed to a higher divine will and plan, which guides the flow of information and ensures that only what is necessary is communicated. In both human and animal relationships, free will plays a significant role in making choices. Consider also that the higher divine will may withhold certain information to respect the free will of the animal or their guardian, allowing them to grow without undue influence. For example, an animal communicator might not be given insight into a specific decision an animal guardian needs to make, such as whether to pursue a particular medical treatment. This could be because the higher divine will wants the guardian to make the decision based on their intuition, research, and consultations with veterinary professionals.

Sometimes, information is not disclosed during a session because the animal and their guardian are not yet ready to receive or process it. The higher divine will may determine that the timing is not right and that revealing certain information could potentially hinder personal growth or create unnecessary confusion. Higher divine will ensures that information is revealed at the most appropriate time for everyone involved. Trusting this guidance is essential. For example, an animal communicator might not receive information about a past traumatic experience of an animal, as the guardian may not be emotionally ready to hear about it. The divine plan may be to disclose this information at a later time. This may also occur when a person is wanting to know when their animal will be transitioning into the afterlife or when the appropriate time is to make that decision on behalf of the animal. The divine will may be that this information shouldn't be shared just then. But a few days or a week later, the higher divine will may support more guidance about this situation.

Remember that higher divine will operates from a place of unconditional love and wisdom, and its guidance is always aligned with the highest good of all involved. Instead of our abilities, we must trust that the higher divine will has a reason for withholding this information, and then focus on providing support and guidance based on the information we do receive. By trusting in the process, we can cultivate a deeper sense of faith in the divine plan and its ability to support us.

Staying committed to your intuitive animal communication practice is a journey that requires patience, dedication, and self-compassion. With the tips in this chapter, you can significantly enhance your skills and deepen your connection with animals. As you continue to grow and develop in your intuitive animal communication journey, remember that you are serving the highest good of the animals and their guardians, and that your commitment to your practice has the potential to make a lasting impact on the lives of those around you.

Journal Reflections

1. How are you applying the techniques you are learning to your everyday life? What are you learning about animals? What are you learning about yourself?
2. What have been some of your successes in your communication with animals? What are some new ways you can celebrate these wins?
3. When and how do you make time for practicing self-care? What are three ways you might try to support yourself better this week? This month?
4. What has been your experience when you have not had energetic boundaries, whether it be with animals or people? What did you learn? When have you felt you've had good boundaries in place and what resulted?

STRENGTHENING AND ELEVATING YOUR ANIMAL RELATIONSHIPS

When we bring pets into our lives, our vision is one of joyful, loving, and supportive relationships. But just like with humans, relationships with pets take work. Often, pet guardians struggle with behavioral issues that they just don't know how to turn around, such as a dog that barks at everything or a pair of cats that won't stop fighting. In this chapter, we will go through some key components for creating strong relationships with your animals: pack dynamics, attention and affection, body language, and instinctual needs.

First, let's meet Valerie and Mugsy, who will help us understand these components through their story. Valerie reached out for my help because her newly adopted two-year-old Lab mix, Mugsy, was very reactive to other people and dogs who came into their house and who they passed while out on walks. Valerie was told that before his adoption, Mugsy was a street dog living with his mother and sibling, who both passed away after being hit by a car. A local man witnessed the accident, caught Mugsy, and brought him to the animal rescue. Valerie felt so sorry for Mugsy and was always telling everyone his story, which would bring her to tears each time she told it.

When I connected with Mugsy, I could feel he was very sensitive and smart. He was like a sponge, always absorbing the energies in his surroundings. In my mind's eye, I saw from his perspective that everyone and everything was a point of energy, and he was assessing each and every one of them with a lot of curiosity, like he had question marks above his head. He was looking to Valerie to help him make sense of this, but he felt that she also had question marks above her head. For example, when people were coming into their home, the visitor was like a big ball of intense, questionable energy for Mugsy. He would feel this and look to Valerie, asking, "I'm uncertain. . . . Do you have this under control?" At that moment, Mugsy would feel Valerie looking back at him with an energy of, "Are you okay? How are you going to respond? What is going on with you?" He would then think, "Okay, I've got questions, you've got questions. I'm not getting the answers I need, so I'm going to take matters into my own paws." Then Mugsy would bark and nip at the visitor. Valerie, nervous and freaked out, would lock Mugsy in a bedroom. This sequence of events led to a negative energetic imprint for Mugsy around visitors, meaning that any time someone came to the door, Mugsy was primed with that imbalanced energy even before the interaction began. The sound of the doorbell alone would trigger his imprinted response.

These imprints were similar on walks. Valerie had no structure when walking Mugsy. She let him lead the way and pull, meaning he was out in front, scouting the area for any incoming threats. Valerie had the question marks above her head energetically, so he would lunge and bark at everyone and everything because he felt she was not providing guidance for what he perceived as potential threats.

Additionally, I felt he was like a lightning rod, attracting energy from all around him. But he would take in so much of this energy that he would short-circuit; he was not properly channeling the energy so he could release it. Valerie was also so attached to Mugsy's traumatic rescue story and the way it made her feel that it was like she was weighing Mugsy down with that sad, heavy energy. Mugsy was showing up to the moment, ready to go, but

Valerie was bringing him back to a place of fear and insecurity. Understanding pack dynamics will help us solve Mugsy and Valerie's issues with walks and visitors.

Pack Dynamics

Pack dynamics play a significant role in the social structure and behavior of many animal species, particularly those that live in groups such as wolves, dogs, lions, and even some birds and primates. Understanding these dynamics and the importance of energetic leadership can help us foster harmonious relationships.

The pack functions as a cohesive unit, working together to hunt, defend territory, raise offspring, and maintain social bonds. No matter what kind of animal you have as a pet, together, you are a pack. Within this structure, individual animals have specific roles and ranks that contribute to the overall success and survival of the group:

- *Pack leader:* When it comes to the animals in your household, you should be the leader. A strong, effective leader can maintain order, ensure the group's safety, and make decisions that benefit the entire pack. Good leaders are calm and confident, provide stability and predictability, and allow the pack members to feel secure and understand their roles within the group.
- *Middle-of-the-pack:* These animals are often the happy-go-lucky ones who adapt easily to changing situations. They often display a high level of social adaptability and are skilled at reading social cues and adjusting their behavior to maintain harmony within the group. They respect the established hierarchy and understand their place within it. To maintain their position within the group, middle-of-the-pack animals often display submissive behavior, such as avoiding eye contact or exposing their belly.

- *Back-of-the-pack:* These members are the sensitive ones who are often the eyes and ears of the pack. They alert the pack when there is something questionable to be aware of and then rely on the pack leader to guide the pack through challenges.

Back to our duo, Valerie and Mugsy. Valerie was not aware that she was not providing the kind of energetic leadership Mugsy needed. When he looked to her for answers and only received more questions and nervous energy, Mugsy was essentially feeling that no one was in charge. This made him feel insecure and like he needed to take defensive action. But that is not who he is by nature, so the whole situation was imbalanced and chaotic for him. I shared with Valerie the idea of the pack hierarchy, and that she needed to first connect to the vision and feeling of being the calm, confident leader. This would allow her to turn those question marks above her head into answers. We are human, and though we are not always going to have the answers, when it comes to our pets, we need to be clear and confident in our role as leader for them.

Valerie also learned that her thoughts and feelings were apparent to Mugsy, so her inner state needed to match what she was asking of him. Asking him to calm down when her own emotions were frenetic with worry was not going to work. I recommended that Valerie make it a practice throughout her day to check in with her feelings and shift toward the positive.

While interacting with Mugsy, I encouraged her to visualize and focus on feeling the experience of the behavior she would like him to understand and experience. I also reminded Valerie to let go of Mugsy's past rescue story as well as the reactivity story she had held on to since he came to live with her. She needed to practice being in the present moment so she could help create new imprints with him, show him that he could rely on her to take the lead, and help him make better, more balanced choices.

Practice: Pivoting toward the Positive

This practice involves shifting your mindset from negative thoughts, feelings, emotions, or experiences to more uplifting ones. Keep in mind, this does not mean ignoring or not honoring negative feelings, because they are valid. But you can honor these feelings while also making an effort to shift them, for your well-being and for the well-being of your pets. By practicing this regularly, you can cultivate a healthier, more optimistic outlook.

1. Commit to focusing on your internal thoughts and emotions. It can help in the beginning to set a periodic alarm on your phone to remind yourself to check in about this.
2. Become aware of any negative thoughts or emotions that arise in your mind. Take a moment to pause and observe them without judgment. Pay attention to the internal dialogue or feelings that contribute to negativity or stress.
3. Recognize that the negative energy is temporary and not a reflection of your true self or reality. Acknowledge your negative thoughts or emotions and allow yourself to feel them without resistance. By accepting these feelings, you can begin to let them go.
4. Imagine the negative energy on one end of a pole and imagine what the thoughts and emotions at the opposite end of the pole would be.
5. Shift your attention to the positive end of the pole and connect with your vision and feeling of experiencing those positive energies. Allow yourself a few moments to simmer in the positive energy as a way of cultivating it further and dissolving the negative.

Awareness of Your Attention and Affection

People are often not aware of their animal's state of mind and energy when they are giving them attention and affection. When an animal is nervous or scared and we comfort them by picking them up, holding them, and offering sweet talk, we nurture that nervous, scared energy. From the animal's point of view, they understand that if they behave in that way, they will receive affection, which further enables their behavior. For example, if your dog is jumping all over someone and you offer them a treat as a distraction, what the dog really thinks is, "Okay, if I jump up on people, I will get a treat!"

This is not isolated to dogs. I had a client whose cats were not getting along, and the person kept picking each of them up, hugging and holding them right after they had a spat. The cats were both getting attention and affection after fighting, and they felt there was a correlation.

The takeaway from this is, be mindful of *when* you are giving attention to your animal, and *what kind* of attention you are giving them. Attention and affection are rewards, so it is best to give them to animals when they are calm. If your animal is behaving in a way you don't like, ignore them and wait for them to relax and stop the behavior before offering praise, petting, treats, and so on. When you reward a calm energy state, they will realize that is what you want from them, and you will have a happier and more well-behaved and balanced animal.

For example, Tammy felt that her Pomeranian, Prince, was the love of her life. She showered Prince with gifts, treats, high-end spa treatments, and the freedom to do whatever he pleased, wherever he pleased. She was always hugging and holding him; she never wanted to be separated from him because she loved him so desperately. But Prince had some behavioral issues because of all this love. He was very protective of Tammy and would not let anyone come close to her. He bit several people and other dogs. He also was not housetrained and was very destructive around the house, tearing up carpets and couches.

When I connected with Prince, I could feel all the love he was receiving. But it felt like all the love was being dumped on him like a bucket of water, and he was drowning in it, confused by what was and was not appropriate behavior. Because Tammy was always and only providing love, she rewarded and enabled all his misbehaviors. Prince misunderstood what was unacceptable, such as biting people who approached her, because she would pick him up, hug him, and cover him with kisses as he reacted. While her intentions were good, her actions misaligned with the highest good for Prince and the others around her. Prince showed me he was intelligent and capable of behaving, but Tammy needed to be more mindful of her emotions, what she was communicating to him, and how her emotions contributed to his actions.

I taught Tammy how to meditate, practice breathwork, and ground herself to balance her emotions. And I showed her new ways to share her love for Prince by creating structure and being more aware of when she was giving him attention and affection. Within a few months, everything shifted for the better. Prince soon began to let people into their personal space without biting or reacting to them and his destructive tendencies disappeared.

Body Language

Becoming more aware of your body language and energy when communicating with animals is crucial for establishing trust and fostering strong connections. Animals are highly sensitive to nonverbal cues, so being conscious of your body language can enhance your ability to communicate effectively with them.

We can go back to Valerie and Mugsy as an example. Valerie was so nervous about how Mugsy might react that she was barely breathing when visitors would come to her house. Every muscle in her body was tense and her arms were crossed. Her visitors, who Valerie had forewarned about Mugsy's reactivity, would enter her home with insecure and guarded body postures. Mugsy picked up on all that buzzy, tense energy and felt that it needed to be addressed and corrected. When Valerie didn't take the lead, Mugsy tried to create order in the situation the only way he knew how, with barking and lunging and other reactive behavior. I encouraged her to shake out her body and loosen up before welcoming people into her home. This alone would send an entirely different—and more positive—message to Mugsy about what to expect, allowing him to shift his response.

When Valerie would walk Mugsy, her body language was tight: her shoulders were crouched down, her head was very low and cautious, and she would tensely wrap his leash around her arm. For dogs, the leash is a main line of energy, so her tension with it was only making him more tense. I recommended that she hold her shoulders back and head high with calm confidence, taking deep breaths as she walked. When she made these adjustments in her body, it made a huge difference in the messages she was communicating with Mugsy, and he began following her guidance with more ease and trust.

Here are some helpful guidelines for how to approach body language when engaging with animals:

- *Adopt an open and relaxed posture:* Avoid crossing your arms or legs, as this can signal defensiveness or tension. Instead, stand or sit with a neutral spine and relaxed shoulders, demonstrating a sense of openness and receptivity to the animal. If needed, shake out your body to help release tension—it really works!
- *Be mindful of eye contact:* While eye contact can be a powerful way to establish a connection with an animal, it's important to be aware of the animal's comfort level. Some animals may find direct eye contact threatening, so be sure to observe their reaction and adjust your gaze accordingly. You can try looking at the animal's forehead or side of their face to maintain a connection without causing discomfort.
- *Take note of your facial expressions:* Animals are sensitive to human facial expressions, so be mindful of the emotions you convey through your face. Maintain a

soft, relaxed expression when communicating with animals, and avoid displaying strong emotions such as anger, frustration, or impatience.

- *Focus on your breath:* Practice slow, deep breathing to help you remain calm and centered during your interactions with animals. This steady rhythm can create a sense of safety and reassurance for the animal, promoting a more open and trusting connection.

In addition to becoming more aware of your body language, you can use an animal's body language to understand their state of mind. Notice their posture, facial expressions, ear positions, tail movement, and other nonverbal cues. Different animals may express fear in unique ways, but some cues are relatively universal. Fearful animals often attempt to make themselves appear smaller by crouching low to the ground or curling up to hide or protect themselves. An animal's ears can also indicate their emotional state. Fearful animals often flatten their ears against their heads. In extreme stress or fear, animals may also exhibit a steady, direct stare. For dogs and some other species, a tail tucked between the legs can be a sign of fear. On the other hand, for cats, an excessively puffed-up tail can also indicate fear. Dogs may yawn a lot as a way of releasing energy when they are anxious or excited. These are just a few examples, so it's essential to observe animals in a variety of situations to understand their specific cues.

Practice: Communicating through Body Language

Practicing communication through body language with animals can help deepen your connection with them and improve your understanding of their needs and emotions. To enhance your nonverbal communication skills, practice with a variety of animals, including different species and individual animals with unique personalities. This will help you become more adept at reading and responding to various body language cues.

1. Begin by observing the animal from afar to understand their body language and the various signals they use to communicate. Pay attention to their posture, facial expressions, tail movements, and other cues.
2. Approach the animal slowly and calmly, stopping when you are in their orbit, but not close enough to touch them. Allow them to observe you and become comfortable with your presence before attempting to interact with them. Some humans naturally have a stronger energetic presence than others, so notice how the animal is responding to your energetic presence—are they backing away or leaning toward you? Just like humans, animals have their personal space boundaries. Be mindful of this and avoid invading an animal's personal space unless they invite you in through their body language.
3. To establish rapport, try mirroring their body language. For example, if an animal is lying down, crouch or sit down to be at their level. This can help the animal feel more comfortable.
4. If the animal seems open, use gentle, nonthreatening gestures. Always move slowly and calmly when approaching an animal, especially if they are unfamiliar with you. For example, extend your hand with your palm facing down and allow the animal to approach and sniff your hand.
5. If an animal seems uncomfortable or displays signs of stress, back off and adjust your approach and body language accordingly. Learn from each interaction and apply your observations to future encounters.

Instinctual Needs

In my animal communication work, I often meet animals who are not feeling calm, confident leadership from their humans. And quite often, the animal's instinctual needs are not being met, which complicates the relationship. Meeting the instinctual needs of your pets is essential for their overall well-being, happiness, and mental health. By catering to these natural instincts, you can create an environment in which your pets can thrive and develop a strong bond and connection with you. This can also help maintain their physical health, ensuring they get the appropriate amount of exercise, nutrition, and rest required for their species, age, and size. Providing outlets for your pets' natural instincts helps keep their minds active and engaged, reducing the risk of boredom, anxiety, and destructive behaviors. Understanding and catering to your pets' instinctual needs deepens the bond between you and your pet, as it demonstrates that you respect and appreciate their unique characteristics. Animals that are unable to express their natural instincts may become stressed, anxious, or frustrated. Meeting their instinctual needs helps to create a more relaxed and content pet.

Physical and Mental Exercise

Ensuring that your pets get enough exercise is important for their overall health as well as to satisfy their instinctual need for activity. This may involve daily walks, playing fetch, or providing toys that encourage physical activity.

Providing appropriate outlets for instinctual behaviors, such as chew toys or scratching posts, will prevent them from damaging your belongings. Look for ways to channel their natural behavior into play, which besides providing exercise is another great way to strengthen your connection with them. Cats in particular have strong predatory instincts that can be engaged through interactive toys that mimic the movements of prey animals, such as feather wands or motorized mice. This can help satisfy their hunting instincts while keeping them mentally and physically active.

Also, it is key to provide mental stimulation through puzzle toys, treat-dispensing toys, or training sessions to engage your pet's mind and satisfy their natural curiosity and problem-solving instincts. It can also provide much-needed mental stimulation as they get older. Social animals, like dogs and cats, have a strong instinct to bond with other animals and humans. Ensure that your pets have ample opportunities for social interaction, whether through playdates with other pets, spending quality time with you, or participating in group training classes or daycare.

Mugsy especially needed a way to channel and release all that energy he was absorbing. I suggested Valerie consider agility, though it didn't have to be formal classes. She could do simple exercises at home like teaching him to hop over a broom at different heights or using cones, boxes, or even pots and pans to create an obstacle course to lead him through. The goals were to remind Mugsy that Valerie was his leader, mentally challenge him, and allow him to showcase his abilities and feel fulfilled.

When it comes to health, nutrition is incredibly important. Feed your pet a balanced and species-appropriate diet to meet their instinctual need for the right nutrients. Consult with your veterinarian to determine the best diet for your pet's age, breed, and health status. Mealtimes should be a calm and uninterrupted time for your pet to enjoy their food without competition or stress. If you use treats or meal toppers, make sure you are offering an appropriate kind and amount. Fresh water should also always be available to your pet.

Structure and Training

It is important to establish clear boundaries and rules for your pets and provide consistent training to meet their instinctual need for structure and guidance. This not only helps to create a harmonious living environment but also allows your pets to feel confident in their place within your household. While traditional training to help with obedience or as a form of mental stimulation is great, there is another aspect of training that is often overlooked, and that's training around your pet's role.

When you think of your pet, what qualities come up? Are they affectionate? Are they aloof? Mentally and energetically assigning your pet a role helps perpetuate that role because they can feel you energetically projecting it on them. If you are stuck feeling your animal is, for example, the scaredy cat or the troublemaker, that is a frequency you are energetically enabling. When we start imagining them in roles that we feel are more appropriate, they will feel this energy and begin to align with it.

In Mugsy's case, Valerie had created this story about him as a defensive, reactive dog. With every episode of his reactivity, it would only feed her attachment to the story. Her story then fed his story, making him more reactive even though I felt it was something he was ready to leave behind. Mugsy wanted to serve the pack, but he didn't know what his role was. He was too busy trying to fill the hole of a leader. Once Valerie stepped in, he was able to fall back into a goofy, loving role.

When we clearly communicate our wishes about the role we want our pets to fill, they will do it. We can do this intuitively, energetically, and telepathically through visualization and feeling them as this new role in our pack and family. Usually, we must also follow through and show them how to assume this new role, but intention and energy form the springboard that guides them toward a more balanced state.

Mantras and Affirmations

To strengthen your animal relationships, you can consider creating your own mantras and affirmations, powerful tools for personal growth, self-improvement, and manifestation. I have found mantras and affirmations to be incredibly helpful for living in the moment and shifting energy from where we have been to where we would like to be. Mantras and affirmations help to remind ourselves of the new energetic vibrations we are aligning with, which is very helpful when it comes to our animals. Repeating a simple mantra or affirmation will help you balance your thoughts, feelings, and emotions in any area of your life. And because animals feel our energy, this will help to improve our connection with them.

Here's a simple guide to help you create your own mantras or affirmations:

1. *Identify your goal or intention:* Start by determining the area of your life you'd like to focus on or the change you'd like to manifest. For animal communication goals, this could be related to calm confidence and leadership or better communication and understanding. You may also wish to focus on other aspects of life, such as your health, relationships, or career.
2. *Write down your thoughts and feelings:* Spend some time with your journal, reflecting on your current thoughts and feelings related to your goal or intention. This will help you gain clarity on the beliefs or patterns that may be holding you back and identify areas where you'd like to create positive change.
3. *Choose your focus:* Select one or two key aspects you'd like to address with your affirmation. This could be a specific goal, a belief you'd like to change, or an emotion you'd like to cultivate.
4. *Begin with a positive, present tense statement:* Your affirmation should focus on what you want to attract or create, rather than what you want to avoid or eliminate. Additionally, your affirmation should be framed in the present tense, as if it's already true. This helps your subconscious mind accept the new belief more easily. If you feel resistance around saying it as if it is already true, you may choose to begin it with, "I am in the process of. . ." For example: "I am in the process of becoming calmer and more balanced."
    - *Make it personal and specific:* Use words and phrases that feel authentic and hold meaning for you. Be specific about what you want to achieve or how you want to feel.

- *Keep it concise and memorable:* Aim for a short sentence or two that you can easily recall and repeat to yourself throughout the day.
- *Add emotion and energy:* Choose words that evoke the feelings you want to experience. This can help increase the impact of your affirmation and make it more effective.

6. *Try it out:* Say your affirmation out loud and notice how it makes you feel. If it doesn't evoke positive emotions or feel authentic to you, revise it until it does.
7. *Practice consistently:* You can repeat it to yourself in the morning, write it down in your journal each day, or use it as a mantra during meditation. Consistent practice will help reinforce the new belief and make it more effective.
8. *Stay patient:* Remember that change takes time, and your affirmation may not yield immediate results. Stay open to the process, trusting that your consistent practice and positive mindset will support your desired outcome.

After Valerie learned about these key components of pet relationships and began applying them, everything shifted for her and Mugsy. Throughout each day she would say her mantra to herself, "I am calm, present, the source for leadership." She would repeat this in her head constantly, and she really began to feel the words. Mugsy was feeling her energy shifting and, because she was more present in the moment, she noticed when he would check in with her for guidance. She made it a regular practice to remind him she had everything handled, and she would step in before his energy imbalanced by mentally challenging him with a training cue or trick. She was also very mindful of rewarding his calm energy state rather than when he was nervous or reactive. She also was doing agility with him out on walks with whatever was around: having him go over and under a bus bench, walk along a narrow picnic seat in the park, or jump from log to log in the forest. Mugsy's instinctual needs were finally being met, and together they were free of the past and creating a new life together.

Journal Reflections

1. What are some challenges you are facing with an animal? What are the feelings you are experiencing around the situation? What opposite, positive feelings would you like to experience?
2. What does a calm, confident, stable energetic partnership with an animal feel like for you? Describe in detail.
3. Where is your attention and energy regarding your animal? Are you focusing on what you don't want or what you do want? How is your attention enabling certain behaviors in your animal? Journal and write a bit focusing on what you do want.
4. How are you feeding the needs of your animals? Are their physical and mental needs being met?

CO-CREATING WITH ANIMALS

When I look back on my life and relationship with my dog Cooper, I appreciate all that we shared and the life we created together. I was not planning on adopting a dog the day I met her, but I listened to my inner guidance and my heart opened up to a profound connection with her. She gave me a new purpose in guiding her through her sensitivities and provided me with love and joy that inspired me to keep going through challenging times of change and growth. She encouraged me to take advantage of the outdoors, hiking and exploring the wilderness where we fed our spirits together. In exchange, I offered her leadership, love, and a belief in her potential to be more calm, confident, and balanced, even in uncomfortable situations. We were always there for each other. She was the one who truly amplified my intuitive animal communication abilities to higher levels and helped me build my confidence to step out into the world and help others like I helped her, making intuitive animal communication into a fulfilling career.

From her puppyhood to the last days of her life, I learned so much from her, which gave me the knowledge to help others who were facing similar situations. There was a symbiotic exchange where she energetically fed me and I her. Every look, every lick, every hike, and every wag elevated my vibration to higher levels where I drew in similar energy that supported the manifestations of all my dreams. My life with her was not just about "owning" a dog; it was an enlightened, shared understanding of what we could create together.

Now that Cooper is in spirit, I still feel her guidance and support. She has brought me to new animals in my life and still helps me support other people and their pets through my work. She is an angel. Because of her, I learned we all have the power to co-create our dreams and desires with our animals. Cooper was and always will be my inspiration, and because of her I was able to bring out the best in her that lit her path to where she is now, an eternal and powerful source of love and light.

Through intuitive connection with our animals, we can understand them at a deeper level, and they also understand us. This deepened relationship continues to build and expand, creating a foundation of clearer communication and understanding upon which to build a balanced and fulfilling life. By strengthening our intuition and awareness of energy, we understand our own needs as well as our animals' needs more clearly. Then we can set intentions for what we would like to create together, visualize it, and connect to what having that feels like. Practicing these feelings consistently is a way to draw the energies of what we desire into our lives with our animals.

As our intuitive and energetic relationship with our animals deepens, we have the opportunity for our communication with and understanding of them to become instinctual and second nature. All that we learn we absorb and apply in every moment. We move through life together as co-creators in mutual love, joy, happiness, and fulfillment.

Animals are a source of unconditional love, and this love is the fuel for the creation of all our desires. In this chapter, we will explore the extraordinary possibilities that arise when you harness your new intuitive animal communication abilities and how you can co-create an amazing life with the animals that share your world. With every step you take, you will realize that this journey isn't just about connecting with animals; it's about discovering a new way of living that resonates with your highest self and reflects your true values.

What Is Co-creation?

Co-creation is a concept rooted in the fields of spirituality and personal development. It refers to the collaborative process through which we can work alongside a higher power, universal consciousness, or the divine to actively manifest our desires and intentions. Co-creation emphasizes the idea that we are not passive observers of our lives, but active participants in shaping our reality. In the context of manifesting and spiritual work, co-creation is based on the belief that our thoughts, feelings, and actions have a direct impact on the circumstances and experiences we attract into our lives.

To engage in co-creation, we must align our desires and intentions with the broader principles of love, compassion, and the highest good. This alignment helps ensure that the process of manifesting is guided by a deeper sense of purpose and serves our personal growth and spiritual development. When we connect with the feelings of love, joy, and whatever makes us feel good, we elevate our energy to the vibrational level that is a match for what we would like to experience. Because animals are such a profound source of love and compassion and possess an innate link to the natural world, our connection with them amplifies the process of manifestation. With awareness of our ability to manifest and co-create with both the Universe and our animals, we can craft the life we desire for and with our animals.

Whenever we experience something that feels good, we are elevated, our mood shifts, we are in a better space, and more great things begin to happen. The challenge is that we are often distracted by other feelings that hold us back, such as doubts, fears, judgments, shame, criticism, and so on. These feelings put stress on our bodies and turn on the stress response system, releasing hormones that are meant to help us fight or flee from potential risks. While the stress response is important in dangerous situations, chronic activation of it puts us at risk for health problems like anxiety, depression, heart disease, sleep issues, and impaired memory, among others. But when we take time regularly to feel love, joy, and happiness, we can physically lower our body's stress response chemicals.

Through meditation and other skills that help redirect our thoughts, we can increase the good chemicals that allow us to feel calm, clear, and open to processing information in a more productive way. This does not have to be a deep meditation (though it could be!).

For a few minutes each day, implement the following meditation to begin your process of co-creation:

Practice: Meditation to Prepare for Co-creation

1. Close your eyes and connect into the feeling of whatever you love—an animal, person, place, experience, food, song, whatever fills you with joy. Bask in that feeling for a few minutes.
2. Bring your attention to something you would like to experience, change, or create. It could be a behavior challenge you are facing with a pet, a challenge in your work or relationships, or something you would like to manifest.
3. Focus with clear intention, and bring that intention into your feeling of love, blending them, and just ***being*** with it. Allow yourself to see and feel what you desire, propelled by the elevated emotions of love and joy.

It really is as simple as just focusing on the good. The trick is making it a habit with consistent practice; even five minutes a day is worth it! Practice this regularly and you will begin to see synchronicities in your life. You will draw to you whatever your intention was focused upon, you will overcome the obstacles that stood in your way, and you will manifest what you desire.

Practice: Embracing Co-creation

Here are some key principles and practices that can help you open your mind to co-creation and its possibilities for a more fulfilling, purposeful, and spiritually enriching life.

1. *Build positive thoughts and emotions:* Our animals are an easy and obvious source of love, gratitude, and joy, so tap into the positive feelings animals offer you. Allow yourself to really feel these positive emotions so they build and expand.
2. *Connect with a higher power:* Whether you identify with a particular deity, universal consciousness, Spirit, your innermost self, Mother Earth, or a more general sense of the divine, take time to nurture your personal spiritual connection. Asking for guidance and support through regular meditation, prayer, or contemplative practices can help deepen your relationship with the higher power that guides your co-creative process.
3. *Surrender control:* Recognize that the process of co-creation involves a balance between personal choice and divine guidance. Trust that a higher power will support you in manifesting your desires in the way that's right for you.
4. *Take inspired action:* While trusting in higher guidance, it is crucial to take proactive steps toward manifesting your desires. Listen to your intuition, and act on the opportunities and ideas that arise in alignment with your intentions.
5. *Practice patience and perseverance:* Co-creation is an ongoing process. Recognize that the manifestation of your desires may unfold in unexpected ways and on a timeline that differs from your expectations.
6. *Cultivate gratitude:* Cultivate an attitude of gratitude for the blessings and lessons that come your way, as well as the guidance and support provided by the higher power with which you are co-creating.

7. *Reflect on your experiences:* Regularly assess the progress of your manifesting journey, and adjust your intentions, thoughts, feelings, and actions as needed. This self-reflection can help you stay aligned with your higher purpose and ensure that your co-creation efforts remain focused and effective.

Cultivating Co-creation with the Support of Animals

While co-creation is certainly something that you can embark on as a solo journey with your higher power, you don't have to do it alone. You already know that animals enrich your life, and they can support your co-creative aims as well.

To embark on this journey, we must first open our hearts and minds to the possibility of a deep, spiritual partnership with these magnificent beings. Recognize that they, too, have desires to love, connect, and feel good. They have unique gifts and wisdom to offer us, and together, we can build a true partnership on mutual respect, trust, and shared energy. We can create a more balanced and fulfilling life for both ourselves and the animals we care for.

To begin, here are some steps and ideas to help you foster this special bond and support your co-creative work:

- *Practice gratitude* by regularly acknowledging the many benefits that animals can bring to your life, such as companionship, stress relief, emotional support, and others.
- *Understand their needs,* the physical, emotional, and mental. Each species and individual animal may have unique requirements for diet, exercise, socialization, and mental stimulation.
- *Establish routines and rituals* to create a sense of stability and comfort for both you and your animals. This can include daily walks, playtime, grooming,

and mealtimes as well as participating in bonding activities that both you and your animals enjoy, such as hiking, agility training, or simply relaxing together. This also means being consistent with the expectations you have for them and approaching every situation with calm, clear intentions.

- *Prioritize self-care and personal growth* such as exercise, healthy eating, and goal-setting to ensure you have the energy and emotional capacity to care for and connect with your animals.
- *Embrace the lessons and wisdom of animals* and apply these insights to your personal growth and self-discovery. Animals can serve as powerful teachers and guides, offering valuable insights into living in the present moment, practicing unconditional love, and embracing the interconnectedness of all life.

Tools for Co-creation

Consistent use of the meditative and intuitive practices in this book will be useful for your journey into co-creation. There are also a couple of specific techniques that have been useful for me and my clients: creative visualization, vision boarding, manifestation boxes, and energetic symbols. Let's take a look.

Creative Visualization

Creative visualization is a mental technique that involves the power of imagination to envision desired outcomes or situations in vivid detail. By repeatedly focusing on these mental images and connecting into how they feel, we can harness the power of our subconscious mind to influence our thoughts, emotions, and actions, ultimately manifesting our desires into reality.

Creative visualization works on the principles of energy and vibrations, which are deeply interconnected with the Law of Attraction. The Law of Attraction asserts that like attracts like, meaning our beliefs have the power to attract corresponding circumstances and experiences into our lives. According to this law, everything in the Universe, including our thoughts and emotions, is made up of energy and vibrates at specific frequencies. When we engage in creative visualization, we are generating mental images and emotions that also carry specific vibrational frequencies. By consistently focusing on these images and emotions, we are effectively tuning in to the energy of the desired outcome, aligning our vibrational frequency with that of our goal. Then, we begin to attract circumstances, opportunities, and experiences that resonate with our goal. This process is what allows creative visualization to influence our reality and manifest our desires.

Our beliefs play a crucial role in the effectiveness of creative visualization. When we believe that our desired outcome is possible and that we are deserving of it, we further strengthen the vibrational alignment. This powerful belief can act as a catalyst, accelerating the manifestation process. When you allow yourself to imagine feeling what it would be like to experience what you desire, you retrain your consciousness to open up to that experience and begin to let go of your conditioned, knee-jerk responses that only get in your way.

There are always going to be situations that challenge us; it's part of being human. But a consistent visualization practice shortens the duration of our triggered responses in any given challenging moment, and it becomes so much easier to bounce back into our desired feelings. Even though the current circumstances may not yet be aligned with our desired outcomes, we can master the current moment and create a new outcome that keeps us on the path. This is empowering, and successes recalibrate our nervous system and prime us for more positive reactions and choices.

You may be thinking, *that's great, but where does my animal come in*? Because our animals are sensitive to energy and look to us for guidance and leadership, they quickly follow our lead and align with our intentions and desired outcome. Creative visualization, therefore, is a great practice to work on behavioral issues with your pets.

For example, if you have a dog that pulls and lunges on the leash while out on walks and you would like to shift your dog's behavior, you can start by connecting with the vision, feeling, and frequency of your dog walking in a calm, surrendering, and peaceful way by your side or behind you. Focusing on the dog not listening to you and your anxiety about walk time only reiterates the feelings of what has been happening. Instead, visualize what you do want, getting clear about your intentions and the energy you would like to experience.

Before the walk, visualize you both walking with peace and ease, imagine what that feels like for you and how your dog feels as they follow your guidance. Check in with yourself to maintain this energy so you are not sending mixed signals. Often, people give up because they are not seeing results fast enough. But it takes as long as it takes. The dog will feel your energetic guidance and begin matching the frequency of your visualization—a perfect example of co-creation.

Vision Boarding

A *vision board,* also known as a dream board or goal board, is a physical, visual representation of your goals, dreams, and aspirations. This can be a powerful tool to help you manifest what you desire for yourself and your animals. It is typically created using a collage of images and words that represent the things you want to attract or achieve in your life. The primary purpose is to serve as a daily reminder of your goals, helping you maintain focus, motivation, and a positive mindset. Vision boards work on the principles of the Law of Attraction, creative visualization, and the power of intention, and have been widely used in personal development and manifestation practices.

Creating a vision board related to animals is a wonderful way to focus on the experiences you wish to manifest. This can include personal goals such as adopting a pet, improving the relationship with your current pet, or pursuing a career working with animals. It can also encompass broader intentions like supporting animal welfare causes, volunteering at shelters, or fostering a deeper connection with wildlife.

Here's how and why vision boards work:

- *Offers visual stimuli:* The human brain is highly responsive to what we see. By creating a tangible representation of your goals, you stimulate your brain's reticular activating system (RAS), which is responsible for filtering information and bringing relevant data to your conscious awareness. A vision board helps to prime your RAS to recognize and prioritize information related to your goals, making it easier for you to spot opportunities and take action.
- *Supports creative visualization:* By regularly focusing on visuals, you engage in creative visualization and align your thoughts, emotions, and energy with your desires.
- *Reinforces positive emotions:* A well-crafted vision board evokes positive emotions like joy, excitement, and gratitude, which help raise your overall vibrational frequency.

I have many clients who have experienced great success with vision boards, including Stephanie and her cat, Willow, who used the process to actualize their dream of living in the countryside, and Tamara, who created a vision board to assist in her health journey along with helping her dog, Bodi, recover from surgery and her cat, Tish, to be less scared and reclusive.

Let's take a look at what worked for Emily, a passionate equestrian, and her horse, Thunder. Emily's dream was to participate in national dressage competitions with Thunder, and she decided to create a vision board. Emily began by gathering images of dressage competitions, champion riders, and horses performing graceful dressage movements. She also included pictures of her and Thunder in their best riding moments, capturing their partnership and shared love for the sport. A picture of a trophy, symbolizing their future success, was front and center. She added words such as *confidence, grace, teamwork,* and *champions* to further solidify her intention. Emily put together the vision board in the stable, where Thunder could be a part of the process. As she added each image and word to the board, she explained her intentions to Thunder. Being a sensitive animal, Thunder nudged her gently throughout the process, seeming to acknowledge and share in her aspirations.

Once the vision board was complete, Emily placed it in the tack room, a place she visited every day before and after riding. Each time she looked at the board, she would visualize herself and Thunder performing gracefully in the competition, feeling the thrill of the crowd, the satisfaction of a well-executed move, and the joy of victory. Over time, Emily found herself practicing more diligently, focusing more on her partnership with Thunder, and even entering local competitions to gain experience. Thunder, too, seemed to respond to Emily's renewed energy and focus, performing better than ever.

Emily and Thunder's journey together evolved as the days turned into months. They shared victories and setbacks, but through it all, their bond strengthened. Emily spent countless hours practicing with Thunder, perfecting their routine, strengthening their communication, and fostering mutual respect. A turning point came when Emily decided to participate in a statewide dressage competition. It was more challenging than the local competitions they had entered before, but Emily felt ready. Each time she felt a wave of doubt, she looked at her vision board and was filled with renewed belief in herself and Thunder.

On the day of the competition, Emily felt a mix of nerves and excitement. As she saddled Thunder, she felt his calm, steady energy, grounding her. She whispered words of encouragement to him, reminiscing about the intentions they had set together, and they performed their routine with grace and precision. The audience watched in awe as they moved as one, their partnership evident in every step. When their routine ended, a wave of applause filled the air, and Emily felt a surge of joy and satisfaction. In the end,

Emily and Thunder won third place—a significant achievement considering it was their first statewide competition. As Emily held the ribbon in her hand, she realized it was the same color as the trophy on her vision board.

Overwhelmed with emotion, Emily thanked Thunder for his unwavering partnership and shared the moment with him. Looking at the ribbon, she realized how their shared vision, represented by the vision board, had materialized. Inspired by this achievement, Emily added the ribbon to her vision board, and they continued to manifest their dream, entering more competitions and winning more ribbons and trophies until, together, they won the national dressage competition.

When Emily placed the national trophy next to the vision board, it was a testament to their journey—a dream visualized, pursued, and finally, achieved. In the end, Emily realized that the power of their success lay not just in her dreams, but in the shared vision and connection between her and Thunder. Together, they had turned a dream into reality. And for Emily, that was the greatest victory of all.

More than just a collection of images and words, the vision board was a tool that helped Emily align her actions with her dream, continually motivating her toward her goal. The vision board acted as a daily reminder that helped align her actions with her aspirations, driving her toward her goal. But more than that, it became a shared symbol of Emily and Thunder's partnership, a testament to their shared dreams and deepening bonds.

Practice: Create a Vision Board

By creating and engaging with a vision board, you harness the power of visual stimuli, creative visualization, positive emotions, and intention to align your energy with your goals and aspirations. Again, this can be for anything, big or small. Maybe it's for your pet to get over his fear of vacuums, maybe it's to create a comfortable, happy atmosphere with new family members, maybe it's to help the both of you through trauma. No matter what, this exercise can help you.

1. ***Clarify your goals*** by journaling to reflect on your dreams and aspirations. Consider various aspects of your life, such as career, relationships, health, personal growth, and hobbies. Write down your goals to create a clear vision of what you want to achieve.
2. ***Gather materials*** including a sturdy board (e.g., corkboard, foam board, or poster board), scissors, glue or tape, and a variety of magazines, newspapers, or printouts with images and phrases that represent your goals. You can also include personal photos, inspirational quotes, and affirmations. There are also vision board digital apps that can be downloaded where you can collect your images.
3. ***Arrange your vision board*** in a way that feels visually appealing and harmonious. You can create categories for different aspects of your life or mix everything together. There's no right or wrong way to do this; just trust your intuition. Once you are happy with the arrangement, secure the images and phrases on your board.
4. ***Display your vision board*** in a prominent location where you will see it daily, such as your bedroom, office, bathroom, living room, or on your computer or phone background if you created a digital version.
5. ***Engage with your vision board daily,*** allowing yourself to experience the positive emotions associated with each goal. Practice creative visualization by imagining yourself achieving your goals and living the life you desire.
6. ***Take inspired action*** by using the motivation and clarity gained from engaging with your vision board to identify opportunities that align with your aspirations. Be open to receiving guidance and trusting your intuition as you work toward achieving your goals.
7. ***Update your vision board*** as you grow and evolve to ensure it continues to represent your current aspirations. This process also serves as an opportunity to celebrate your achievements, acknowledge your progress, and express gratitude for the blessings and opportunities that come your way.

Manifestation Boxes

A *manifestation box*, also known as a creation box, intention box, or wish box, is a container in which you place written intentions, images, or objects that represent your goals and desires. It is the same concept as a vision board, serving as a tangible representation of your aspirations, but it can be useful if you have a more limited space to work with or can't put up a vision board where you want to.

Here's how and why manifestation boxes work:

1. *Helps clarify and set intentions:* Writing down your goals is a powerful act that helps you get clear about your desires and then communicate them to the Universe.
2. *Focuses attention:* Regularly interacting with your manifestation box, such as adding new intentions or reviewing existing ones, helps maintain focus on your goals and keeps them at the forefront of your mind.
3. *Allows the practice of surrendering control:* Placing your intentions in the manifestation box symbolizes entrusting your desires to the Universe, allowing it to work on your behalf to manifest your goals. This act of surrender helps to release any resistance or attachment to specific outcomes and encourages the Universe to deliver your desires in the best possible way.

The manifestation box, with its potent symbolism and energetic properties, serves as a beacon for your dreams. Each intention placed within it is like a message sent out into the Universe, declaring your readiness to receive the experiences, relationships, and opportunities that align with your highest good. Your animal can contribute their unique vibration to the process, amplifying the power of your intentions and adding an extra layer of love, trust, and shared purpose. Their presence serves as a beautiful reminder of the deep bond you share and the power of unconditional love in the manifestation process.

To see how this can work, let's take a look at Janet, an ardent animal lover, and Charlie, her energetic border collie. Janet's goal was to establish a holistic wellness routine for herself and Charlie. She wanted them both to be in peak physical and mental health as well as to deepen their bond through shared activities. She started by selecting a beautiful wooden box that she felt a strong connection with, and she decorated it with painted paw prints and symbols of wellness, like apples and yoga poses.

To set her intentions, Janet sat down with Charlie by her side, his head resting on her lap. She took a deep breath, petted Charlie, and wrote down her wishes on small slips of paper. "Charlie and I enjoy a healthy diet," she wrote, and "Charlie and I engage in daily exercise." She wrote each intention in the present tense, affirming it as if it were already a reality. She took a moment to visualize every intention, to feel the joy of sharing a nutritious meal with Charlie, or the satisfaction of a good run. Each time, she noticed Charlie's tail wagging, his eyes meeting hers in silent understanding.

Once she'd written all her intentions, she folded the pieces of paper and placed them in the box. Each time she added a slip, she gave Charlie a gentle pat, acknowledging his role in this process. She placed the box on a shelf in her living room, a place where she and Charlie spent a lot of time together. Every day, she'd look at the box, take a moment to visualize her intentions, and thank the Universe.

Over the next few weeks and months, Janet noticed changes. She found herself more motivated to cook healthy meals, and Charlie seemed more eager for their daily walks. Their bond grew stronger and their shared activities more frequent as they co-created the life they wanted together.

With a manifestation box, you plant the seeds for change. Then, what almost always happens is, after six months or a year, you go back and look in your box and see that you have manifested everything you put inside it. For me, if any intentions didn't come through, I take some time to understand why. Often, we initially think we want or need to experience something when it's not in our best interest or supporting our highest good. Also, our intentions can manifest in unexpected ways, so remember to open up your interpretation.

Practice: Create a Manifestation Box

A manifestation box helps you plant the seeds for co-creation with your animal and serves as a physical reminder of your goals. Maybe you would like to discover a new shared experience together, such as an activity class or a new place to explore together. Or maybe you would like to manifest a new house with more space and a beautiful yard. You may even desire improved health and vitality for you both, or calm and peaceful interactions with other people and animals.

1. *Choose a container* that resonates with you and feels special. This could be a wooden box, a decorated shoebox, or even a beautiful glass jar.
2. *Personalize the box* by decorating it with images, symbols, or words that represent your goals or that simply inspire you. You can paint, draw, or attach printed images and affirmations.
3. *Craft your intentions* on small pieces of paper or notecards. Be specific and positive, and write them in the present tense, as if they have already happened. For example, "My animals and I intuitively communicate and understand each other more deeply with each day," or "I successfully open my own animal rescue organization." You can also add images or small objects that represent your desires, such as a photo of a specific dog breed you wish to adopt or one of a location you would like to live.
4. *Place your intentions in the box.* If possible, involve your animal or turn it into a little ritual.
5. *Engage with your box* by putting it where you will see it daily. Spend a few moments each day focusing on your intentions and experiencing the positive emotions associated with achieving your goals.

6. ***Take inspired action*** by using the motivation and clarity gained from engaging with your manifestation box to identify and pursue opportunities that align with your animal-focused aspirations.
7. ***Continue the process.*** As you witness the manifestation of your desires, remember to express gratitude for the blessings and opportunities that come your way. After you have achieved the goals, you can burn the paper to release the energy into the atmosphere (safely of course!), recycle the paper, or do any other special ritual you feel guided to complete. Then add new intentions and continue the process.

Energetic Symbols

A *symbol* is anything that represents something else, and in this case, it is something material standing in for something abstract. Creating a symbol to energetically charge and represent a goal is an effective technique for manifestation. While a vision board or manifestation box takes up space and tends to stay in one place once you make it, a symbol is something you can take with you anywhere. By associating a specific symbol with a particular intention, you create a mental and energetic link that serves as a reminder and reinforces your commitment to achieving that goal. This technique is rooted in the principles of focused attention, intention setting, and the power of symbolism.

Let's consider the story of Mark and his parrot, Sunny. Mark's dream was to create a wildlife rescue and rehabilitation center. Having cared for Sunny since she was a chick, Mark had developed a profound respect and love for animals, and he wanted to extend his care to more wildlife in need. To help manifest this, Mark decided to create a symbol to represent his goal. He chose the image of a tree to represent growth, shelter, and life. Mark drew this on a piece of paper, adding details like a bird perched on a branch and a small animal resting at the tree's roots. He also wrote his intention beneath the tree: "A safe haven for wildlife to heal and grow."

To involve Sunny in the process, Mark explained his dream to her as he drew the symbol. With her bright, curious eyes, Sunny watched Mark, tilting her head and softly chirping as if she understood. Once the symbol was complete, Mark placed it on his desk, a place he visited daily. He also made a smaller version of the symbol, which he carried in his wallet. Each time he saw the symbol, he took a moment to visualize his dream sanctuary, feeling the satisfaction of helping wildlife and the joy of seeing them recover. Mark found that the symbol served as a powerful reminder of his dream. He started researching wildlife rescue centers, reaching out to experts, and even volunteering at a local sanctuary to gain experience. His actions started aligning with his dream, and he could feel himself moving closer to his goal. As for Sunny, she seemed to sense Mark's growing commitment. She became more attentive when Mark spoke about his dream, often flying over to perch on his shoulder as he researched or planned.

Several months passed, and Mark had raised enough funds to buy a piece of land that was ideal for his wildlife sanctuary. He had been granted the necessary permits and started building the first structures. Mark was thrilled, and so was Sunny. Every time he left for the site, she would let out a series of excited chirps, as if she was cheering him on. One day, while Mark was out at the construction site, he noticed a small bird with a broken wing. Carefully, he picked up the bird and brought it home. He set up a temporary place for it to stay, fed it, and started nursing it back to health. All the while, Sunny watched attentively from her perch. It was as if the small bird was the living embodiment of Mark's symbol.

Once the bird began to recover, Mark decided to name it Hope, symbolizing his aspirations for the sanctuary. As Hope thrived under Mark's care, so did the sanctuary. More funds were raised, more people got involved, and slowly, the sanctuary was ready to welcome its first official residents. When the opening day arrived, he looked at the symbol he had drawn months ago. He realized how each element of it was coming alive: the tree had turned into the sanctuary, and the bird on the branch was Hope. As he was lost in these thoughts, Sunny flew up and perched on his shoulder, softly chirping in his ear. It was a moment of profound connection between them, strengthening Mark's resolve to continue on this path he had chosen. He knew then that his symbol had not only helped manifest his dream but had also deepened his connection with Sunny.

From that day forward, Mark's wildlife rescue and rehabilitation center thrived. It became a haven for animals, just like he had envisioned. And Sunny? She was always there, right by Mark's side, supporting him in her unique way. This experience was a powerful testament to the profound bonds that exist between humans and animals. By focusing on his goal, embracing the symbol, and including Sunny in the process, Mark successfully manifested his dream. His story served as an inspiration to others, proving that with a clear intention, unwavering focus, and the support of our animal companions, our dreams can indeed become reality.

Practice: Create an Energetic Symbol

Creating a symbol to manifest your intentions with your animals is a beautiful way to solidify your goals and make your dreams more tangible. For example, you may choose a figure eight/infinity symbol to represent your intuitive connection with an animal. Or a crown to represent your desired role as a calm, confident leader for your pets. You may even visualize a beautiful peace symbol to represent the calm relationship you are creating together, free of stress and tension.

1. ***Identify your intention*** by asking, "What would I like to manifest?" The goal should be clear, positive, and aligned with your values and desires. For example, "I would like to improve my communication skills with animals."
2. ***Create a symbol*** that represents your goal. This can be an abstract shape, a meaningful symbol, or a combination of both. The symbol should resonate with you on a personal level and evoke positive emotions associated with your goal. Keep the design simple enough that you can draw, paint, or

even make a digital image of it. It could take up a whole wall in your home or you could inscribe it on a necklace. Make it colorful or monochrome, detailed or simple, depending on what resonates with you. As you create it, visualize your intention, and feel as if it's already a reality. Involve your animal in the process however feels right for you.

3. ***Charge the symbol*** by finding a quiet space where you can focus without distractions. Using the physical version or a visual representation of your symbol, spend a few minutes meditating on your goal and the positive emotions associated with achieving it, such as happiness, satisfaction, or fulfillment. As you do this, imagine infusing the symbol with the energy of these emotions and your intention.
4. ***Engage with the symbol*** every time you see it. Take a moment to visualize your intention as already manifested. Feel the emotions associated with achieving your goal. If possible, involve your pet in this process, cuddling with them or speaking to them about your shared intention.
5. ***Take inspired action***, for example, if you intend to improve communication with your pet, you might be inspired to spend more quality time with them, learn more about their behavior, or seek professional advice.
6. ***Recognize changes*** as signs that you are moving closer to your goal and remember to express gratitude for the progress you make and the opportunities that come your way.

Journal Reflections

1. Co-creation begins with your intentions. Reflect on the relationship and shared experiences you wish to manifest with your animal. What are your biggest dreams and aspirations for your shared life? How do you envision your day-to-day interactions? What

feelings or emotions do you want to experience in your relationship? What are your specific goals, such as training achievements, travel experiences, health aspirations, or lifestyle changes?

2. Think of a time when you used creative visualization—even if you didn't call it that, you have probably spent time imagining how a certain situation might turn out or daydreaming about something you wanted to happen. You might even think of a time when you had a lot of anxiety over something and worried about a bad result. What happened? How do you think your creative visualization might have played into the outcome?
3. Which of the practices—creating a vision board, manifestation box, or energetic symbol—appeals to you? Why? What about it excites you? What about it makes you apprehensive?
4. How might you involve your pet in the process of creating a vision board, manifestation box, or energetic symbol? In what ways might this activity deepen your connection with your pet and align your energies?

AFTERWORD

As you turn the last pages of this book, you find yourself not at the end of a journey but rather at the start of an extraordinary new adventure. Together, we have navigated uncharted territories, explored profound truths, and embarked on a path of understanding that has brought us closer to our animal companions and, ultimately, to ourselves. I hope this book has served as a compass, guiding you through an introduction to the beautiful language that exists beyond words, a language of intuition, emotion, energy, and spirit that bridges the gap between species and deepens your connection to the natural world.

This exploration has not been without its challenges. It has asked you to question your preconceptions, step out of your comfort zone, and open your mind to possibilities that may have previously seemed far-fetched or even impossible. But with each step, you have gained a deeper appreciation for the intricacy of the animal world, and for your place within it. You have come to understand that every creature, no matter how big or small, holds a unique and vital role.

More than understanding, this journey has been about transformation and developing a greater capacity for empathy, compassion, and love. These qualities, once kindled, have the power to radiate beyond your individual life, touching others and ultimately contributing to a global shift in consciousness. This transformation is the true gift of intuitive animal communication. As we grow in our understanding and appreciation of animals, we simultaneously nurture a more profound respect for our shared home, the earth. This symbiotic relationship is a powerful force for healing—both for ourselves and our planet.

This book offers a foundation for a lifelong practice, and it will always be here for you to explore. But I'd also like you to see it as an invitation to dive deeper, to venture further, to continually expand your horizons and unique talents. The insights and techniques within these pages provide tools to connect with animals. But the keys to the kingdom of animal communication lie within you. Your intuition, empathy, and love for animals—these are the magic ingredients that will fuel your journey forward.

The truth is, the wisdom we can gain from animals is immeasurable. They guide us toward truths that we may otherwise overlook. Their perspectives, uninhibited by the complexities of human consciousness, offer a unique window into the essence of life itself. Through their eyes, we can glimpse the world in its raw, unfiltered beauty, unburdened by the cultural narratives and societal expectations that often cloud our human perception.

Going forward, I invite you to share what you've discovered with the world—you have the chance to make a difference. As intuitive communicators, we have an opportunity to serve as ambassadors for the animal kingdom. We can advocate for their rights, promote their welfare, and work to create a world where every animal is treated with the respect and kindness they deserve. We can share their stories, spread their wisdom, and amplify their voices.

Our journey into the world of intuitive animal communication is a path of service, love, and growth. It is a journey that calls us to be better, do better, and love better. I encourage you to trust your intuition, follow your heart, and embrace the joy of discovery. Remember, the goal is not perfection, but connection.

So, here's to the animals—our teachers, guides, and friends. Here's to the future, and to the endless possibilities that await us. Here's to us, and the future we are shaping together. May this practice inspire you, challenge you, and enrich your life in ways you never imagined. May your journey continue to be filled with love, understanding, and countless moments of connection. And may you always remember, in the grand tapestry of life, your thread is woven together with that of every animal you meet.

REFERENCES

Chapter 2

1. Sakurai, J. J., and Jim Napolitano. *Modern Quantum Mechanics*. Cambridge, UK: Cambridge University Press, 2017.

Chapter 3

1. Craig, Gary. *The EFT Manual*. Fulton, CA: Energy Psychology Press, 2011.
2. Goleman, Daniel. *Emotional Intelligence: Why It Can Matter More Than IQ*. New York, NY: Bantam Books, 1995.
3. Fredrickson, B. L., et al. "Open hearts build lives: Positive emotions, induced through loving-kindness meditation, build consequential personal resources." *Journal of Personality and Social Psychology*, 95(5) (2008): 1045–1062. https://doi.org/10.1037/a0013262
4. Helminski, Kabir. *The Knowing Heart: A Sufi Path of Transformation*. Boulder, CO: Shambhala Publications, 1992.
5. Judith, Anodea. *Eastern Body, Western Mind: Psychology and the Chakra System as a Path to the Self*. New York, NY: Celestial Arts, 2004.
6. Sapolsky, Robert M. *Why Zebras Don't Get Ulcers: The Acclaimed Guide to Stress, Stress-Related Diseases, and Coping*. New York, NY: W.H. Freeman and Company,1994.
7. Nhat Hanh, Thich. *The Miracle of Mindfulness: An Introduction to the Practice of Meditation*. Boston, MA: Beacon Press, 1999.

Chapter 4

1. "The importance of staying hydrated." *Harvard Health Letter* published by Harvard Health Publishing, June 18, 2015. https://www.health.harvard.edu/staying-healthy/the-importance-of-staying-hydrated

Chapter 5

1. Radin, Dean. *The Conscious Universe: The Scientific Truth of Psychic Phenomena*. New York, NY: HarperCollins, 1997.

Chapter 6

1. Campbell, Joseph. *The Way of the Animal Powers*, vol. 1, bk 1 of *Historical Atlas of World Mythology*. New York, NY: HarperCollins, 1983.
2. Cooper, J. C. *Symbolic and Mythological Animals*. New York, NY: HarperCollins, 1992.

3. Morrison, Lesley. *The Healing Wisdom of Birds: An Everyday Guide to Their Spiritual Songs & Symbolism*. Woodbury, MN: Llewellyn Publications, 2011.

Chapter 8

1. Campbell, Joseph. *The Way of the Animal Powers*, vol. 1, bk 1 of *Historical Atlas of World Mythology*. New York, NY: HarperCollins, 1983.

Chapter 9

1. Brennan, Barbara. A. *Hands of Light: A Guide to Healing Through the Human Energy Field*. New York, NY: Bantam, 1988.
2. Krippner, Stanley, and Daniel Rubin. *The Kirlian Aura: Photographing the Galaxies of Life*. New York, NY: Anchor Books, 1974.
3. Leadbeater, C. W., *The Chakras*. London, UK: Theosophical Publishing House, 1974.
4. Myss, Caroline. *Anatomy of the Spirit: The Seven Stages of Power and Healing*. New York, NY: Harmony, 1996.

Chapter 10

1. Suzuki, Shunryu. *Zen Mind, Beginner's Mind: Informal Talks on Zen Meditation and Practice*. Boulder, CO: Shambhala Publications: 1970.
2. van Dernoot Lipsky, Laura, and Connie Burk. *Trauma Stewardship: An Everyday Guide to Caring for Self While Caring for Others*. Oakland, CA: Berrett-Koehler Publishers, 2009.

Chapter 11

1. Millan, Cesar, and Melissa Jo Peltier. *Cesar's Way: The Natural, Everyday Guide to Understanding & Correcting Common Dog Problems*. New York, NY: Three Rivers Press, 2006.

Chapter 12

1. Bear, Mark F., Barry W. Connors, and Michael A. Paradiso. *Neuroscience: Exploring the Brain*, 4th ed. Philadelphia, PA: Lippincott Williams & Wilkins, 2015.
2. Dispenza, Joe. *You Are the Placebo: Making Your Mind Matter*. Carlsbad, CA: Hay House, 2014.
3. Dyer, Wayne W. *The Power of Intention: Learning to Co-create Your World Your Way*. Carlsbad, CA: Hay House, 2004.
4. Hicks, Esther, and Jerry Hicks. *The Law of Attraction: The Basics of the Teachings of Abraham*. Carlsbad, CA: Hay House, 2006.
5. Mayo Clinic Staff. "Chronic stress puts your health at risk." Mayo Clinic. December 14, 2023. https://www.mayoclinic.org/healthy-lifestyle/stress-management/in-depth/stress/art-20046037

ACKNOWLEDGMENTS

I would like to first and foremost thank my family for their limitless love, encouragement, and support: my parents, John and Mary Fran; my siblings, Judy, John, Kevin, Marie, Mary, Dan, Kathy, and Marron; my partner, Donald; and all of my animals who have changed my life and inspired me: Cooper, Peanut, Harlow, Dasher, Atticus, Buddy, Luca, Mr. Snowbie, Lady Snowbie, Monty, Dolly, Corky, Rocky, Cory, and Gatsby.

I am so profoundly grateful to Colette Baron-Reid. Your wisdom, guidance, and unwavering faith in my abilities have continually inspired me. Your mentorship has been more than just professional advice; it's been a compass guiding me through this journey. Without your invaluable support and encouragement, this work would not be what it is today. Thank you for your generosity of spirit and for the wealth of light you have shared.

I extend my love and heartfelt gratitude to Cesar Millan for believing in me, inspiring me, and for creating the opportunity to connect with, help, and serve so many wonderful people and animals from all over the world.

I would also like to send many big thanks to: John Narun, Linda Howe, Dawn Silver, Chris Walker, Sonia Choquette, Christy Whitman, C. A. Brooks, Laura Lynne Jackson, Debra Katz, Joan Ranquet, Krista and Clay Rimpa, Debra Silverman, Brigit Esselmont, Janet Jackson, Dana Trujillo, and Colleen Steckloff. And to all the people and animals I have communicated and connected with all over the world who I learn from and who inspire me every single day.

Additionally, I would like to express my deepest gratitude to Reid Tracy and Hay House for their belief in this project and for their professional guidance. I am so grateful to Audra Figgins, Anna Cooperberg, and Nirmala Nataraj for their help and support.

Lastly, to the readers and their animals, thank you for picking up this book and embarking on this journey with me. I hope you find as much joy in reading it as I found in writing it.

ABOUT THE AUTHOR

Michael R. Burke is an animal communicator, behavior consultant, and intuitive coach. His mission is to help people better understand and communicate with their animals, to overcome behavioral challenges, and to create longer-lasting, more loving bonds between people and their animal families. Michael communicates with animals intuitively, energetically, and telepathically, living and in the afterlife, and he teaches others to do the same. Michael provides guidance in answering questions, finding solutions, and providing people and animals with peace, harmony, confidence, and clarity.

Michael holds a B.A. in environmental studies from the University of Colorado at Boulder. He is also a certified life coach, meditation teacher, certified dog trainer, and animal behaviorist. He has worked alongside Cesar Millan at his Dog Psychology Center in Southern California as a dog trainer and behaviorist and as the meditation director of the Training Cesar's Way programs.

Michael's unique approach combines his background as an intuitive, meditation teacher, and coach with his experience in dog psychology, animal behavior and training, energy healing, and personal development.

Hay House Titles of Related Interest

YOU CAN HEAL YOUR LIFE, the movie
starring Louise Hay & Friends
(available as an online streaming video)
www.hayhouse.com/louise-movie

THE SHIFT, the movie,
starring Dr. Wayne W. Dyer
(available as an online streaming video)
www.hayhouse.com/the-shift-movie

***

ANIMAL COMMUNICATION MADE EASY: Strengthen Your Bond and Deepen Your Connection with Animals by Pea Horsley

ANIMAL MAGIC: The Extraordinary Proof of Our Pets' Intuition and Unconditional Love for Us by Gordon Smith

DOGS AND CATS HAVE SOULS TOO: Incredible True Stories of Pets Who Heal, Protect and Communicate by Jenny Smedley

ANIMAL APOTHECARY: A 44-Card Oracle Deck and Guidebook for Manifestation & Fulfillment by Cara Elizabeth

All of the above are available at your local bookstore or may be ordered by visiting:

Hay House UK: www.hayhouse.co.uk
Hay House USA: www.hayhouse.com®
Hay House Australia: www.hayhouse.com.au
Hay House India: www.hayhouse.co.in

All of the above are available at your local bookstore,
or may be ordered by contacting Hay House.

***

'The gateways to wisdom and knowledge are always open.'

Louise Hay